The West without Water

The West without Water

WHAT PAST FLOODS, DROUGHTS, AND OTHER
CLIMATIC CLUES TELL US ABOUT TOMORROW

B. Lynn Ingram and Frances Malamud-Roam

Foreword by Sandra L. Postel

UNIVERSITY OF CALIFORNIA PRESS

BERKELEY LOS ANGELES LONDON

University of California Press, one of the most distinguished university presses in the United States, enriches lives around the world by advancing scholarship in the humanities, social sciences, and natural sciences. Its activities are supported by the UC Press Foundation and by philanthropic contributions from individuals and institutions. For more information, visit www.ucpress.edu.

University of California Press
Berkeley and Los Angeles, California

University of California Press, Ltd.
London, England

Library of Congress Cataloging-in-Publication Data

Cataloging-in-Publication Data is on
file at the Library of Congress.
 ISBN 978–0–520–26855–5 (cloth : alk. paper)
 ISBN 978–0–520–94580–9 (ebook)

Manufactured in the United States of America

21 20 19 18 17 16 15 14 13
10 9 8 7 6 5 4 3 2 1

The paper used in this publication meets the minimum requirements of ANSI/
NISO Z39.48–1992 (R 2002) (*Permanence of Paper*).

For my loving family: Don, Daniel, Aaron, and Gene, and in memory of my father, Dr. Gerald Ingram

B. Lynn Ingram

For Anne

Frances Malamud-Roam

CONTENTS

ILLUSTRATIONS

FOREWORD

There is little doubt that humanity is in for turbulent times when it comes to water. Rising human demands against a finite supply are draining rivers, shrinking lakes, and depleting aquifers. In a world of seven billion people and growing, competition for water is intensifying to quench our thirst, grow our food, generate electricity, and manufacture all manner of consumer goods from cars to computers to cotton shirts.

On top of these demand-driven trends, the last century and a half of greenhouse-gas emissions and the concomitant rise in global temperatures are fundamentally altering the cycling of water between the sea, the atmosphere, and the land. Climate scientists warn of more extreme floods and droughts and of changing precipitation patterns that will generally make dry areas drier and wet areas wetter. In terms of water, the natural variability of the recent past will not be a reliable guide to the future. We have moved outside the bounds of "normal" to some new normal yet to be understood.

The recent decade of drought in the southwestern United States has thrown a spotlight on this region's vulnerability to the changes that lie ahead. From 2000 to 2009, the Colorado River exhibited the lowest ten-year-running-average flow of any ten-year period in the past century. During that time, the capacity of giant Lake Mead—which stores water for nearly 30 million people and vast areas of farmland—fell from 96 percent to 43 percent. Climate scientists at the Scripps Institution of Oceanography have estimated that, within a decade, there is a 50 percent chance that Lake Mead will drop below the reservoir's outlets.

It is sobering to realize that this current southwestern drought pales (at least so far) in comparison to the "mega droughts" that have occurred multiple times in the past—and that could occur again in the future. The warm "medieval drought" during the middle of the twelfth century lasted two decades and was more severe and widespread than any other in the Southwest over the past 1,200 years. It almost certainly factored into the decline and ultimate collapse of the Hohokam, an enterprising people who settled and farmed in the valleys of the Salt, Gila, Verde, and other rivers of southern Arizona. The Hohokam thrived for a thousand years—from 450 to 1450—and then disappeared as a distinct culture.

Today, in our technologically sophisticated world, it is easy to believe we are immune to such an outcome. But we have not been seriously tested. For the past century and a half, humanity has enjoyed a relatively benevolent climate. During this time, we built big dams to tame the earth's rivers, diverted flows from one river basin to another, drained wetlands for farming, and built oasis cities in the desert. We have so successfully masked aridity that we have become imbued with a false sense of security about our water future.

In this engaging and enlightening book, B. Lynn Ingram and Frances Malamud-Roam peer deep into the past through the lens of paleoclimatology to assemble evidence that can inform how citizens and leaders prepare for the new climatic regimes, including more intense droughts and floods, that almost certainly lie ahead for the western United States. The insights they uncover through this climate detective work lead them to a sense of urgency about taking action now to prepare for more erratic and extreme future conditions.

We may have other lessons to learn from the past as well. In order for our modern culture to enjoy a millennium of prosperity in the West, as the Hohokam did, we may need to embrace aridity and water's tightening limits and apply ingenuity and creativity to successfully live within those limits, rather than continue to mask and deny them.

Sandra L. Postel
Director, Global Water Policy Project

I love a sunburnt country,
A land of sweeping plains,
Of rugged mountain ranges,
Of droughts and flooding rains.

DOROTHEA MACKELLAR,
"My Country"

A noise woke me in the middle of the night. I sat bolt upright in bed, my heart pounding, but heard only the sound of rain pelting the roof. Peering out my bedroom window, I saw a flashlight bobbing in the blackness of the backyard. My father, wearing a yellow raincoat, was pointing the flashlight toward the edge of our yard. The rain had been heavy on and off for weeks, soaking the ground, causing erosion, and undercutting our yard and those of our neighbors. Where once we had an ample yard, now only about fifteen feet of grass separated our house from the precipitous edge of the canyon. My father was discovering that one of our trees had vanished with the latest chunk of our property to disappear down the slope.

We lived on a steep mountain ridge in Santa Barbara, a small, beautiful city in Southern California nestled between the Santa Ynez Mountains and the Pacific Ocean, about sixty miles west of Los Angeles. By the end of that wet winter of 1969, several homes on our street had been largely destroyed. The house to our right balanced precariously on the edge of a newly formed cliff face, the former backyard entirely gone. Those neighbors, in an effort to save their home, shored up the eroding property by piling truckloads of large gray boulders onto the slope below. But nature finally won out, and they were forced to leave. The neighbors on the other side also lost most of their backyard; only their concrete swimming pool remained, fully intact, surreally suspended over the canyon.

Although our home was spared that year, the torrential rains saturated hillsides and triggered massive landslides and mudflows all across Southern California. Roads were transformed into rivers and flat lowlands into muddy lakes. The river flows in Santa Barbara doubled their normal levels. More than 100 people were killed, and hundreds of homes were destroyed. Southern California was declared a federal disaster area, and the National Guard was brought in to help with cleanup and restoration. As an eight-year-old, I had no way of understanding these events.

Pouring rain is a common occurrence during Southern California winters, but so are sunshine and heat. This hot-weather climate extreme is of greater long-term concern, because it is the cause of perennial scarcity of freshwater in California and the Southwest. Many of my childhood summers were spent in central Arizona on my grandparent's ranch in the Sonoran Desert south of Phoenix, with more climate lessons. Central Arizona and Southern California are similar in climate—different ends of the same desert. The surreal temperature contrast between the cool of the air-conditioned houses and cars and the oven-like heat outdoors is still a vivid memory. Summer temperatures in the central Arizona desert are routinely among the hottest in the nation, commonly reaching 115–20°F, and, though I noticed the cotton and alfalfa fields, I never questioned how these crops could grow in a desert. Nor did I wonder how water could flow freely from faucets in the Arizona desert, just as it did in arid Southern California.

Occasionally, the limitations of living in a desert become more evident. I clearly remember the drought of 1976–77, which was the year that California—and much of the western United States—had virtually no rain. As a teenager, I was struck by the realization that water is a finite resource, one that can actually run out. My friends, my family, and I had always used water liberally with little thought about supply. But in 1977, suddenly every drop counted. We were prohibited from washing cars, watering lawns, and taking baths or long showers. Of course these "sacrifices" paled in comparison to the far harsher impacts we heard about on the news: farmers with less water, ski areas with no snow, and forests drying and burning.

A decade later, drought returned to California, this time one of the longest on record, lasting from 1987 to 1992. I was then a graduate student at Stanford University, ready to begin researching California's climatic past. As an adult, I wanted to understand the forces that led to the 1969 rainstorms and the 1977 "year with no rain." I had been fortunate to grow up during two decades when California experienced a dependable abundance of precipitation. In my third decade, however, we were seeing only half the normal amount of rain and snow each year. Which was the "real" California climate, I wondered. Was there such a thing as a "normal" climate for California and the West?

The goal of my graduate research was to look at how California's climate, especially its rainfall, had varied over the hundreds and thousands of years before records were kept. I focused on the past 11,000 years, the Holocene period, which extends from the end of the last ice age—when the enormous North American glaciers melted—to the present. I focused my research on the San Francisco Bay, an estuary in the heart of one of the largest population centers of the West Coast. The bay is an excellent place to search for clues about California's past climates, because at its bottom are layers of sediment containing subtle chemical signals that reflect the strengthening and weakening of the flow of California's two major rivers, the Sacramento and the San Joaquin.

San Francisco Bay and its inland delta receive river water that begins as rain and snow in locations that span almost half the state. During years with high rainfall, the rivers bring abundant water to the bay, causing the salt content of the water to decrease. During the dry years, the rivers bring less freshwater to the estuary, and the bay becomes saltier—more like the Pacific Ocean waters that meet those of the bay at the Golden Gate.

In my research, I came to focus on the chemistry of fossil shells from sediment cores taken from beneath San Francisco Bay. The chemical signals embedded in those cores allowed me to estimate how the amount of river water entering the bay had changed over many centuries. I found that the river flow fed by rainfall in California has undergone major swings over the past 5,000 years. The alternating wet and dry periods lasted decades or even centuries—much longer than the six-year drought of 1987–92. To check these findings, I worked with my colleagues Roger Byrne, Frances Malamud-Roam, and Scott Starratt on studies of marshlands surrounding San Francisco Bay to examine how these long climate swings affected the marsh ecosystems. Our studies showed that marsh vegetation comprised more species adapted to saltier water during the dry periods and more species adapted to fresher water during wet periods. And, like the sediment fossil records, these swings in the marsh ecosystems lasted decades or longer.

My family had spent a year in southeast Australia, where I later returned for two sabbaticals. The landscapes and the climate there in many ways mirror those in the western United States, with water just as sporadic and extreme. Yet only when I was well into my research career did I realize that these places I loved—California, the American Southwest, and Australia—were vulnerable and growing more so every decade. Long periods of drought punctuated by catastrophic floods over the past millennia are a part of the climate history of Australia, just like that of the American West. These regions are also likely to share a similar future, according to climate models. As the global climate warms, the American West and Australia will become drier, with longer and deeper droughts interspersed with catastrophic floods.

As so eloquently expressed by Australian poet Dorothea Mackellar in "My Country," written in 1911 at the end of a long Australian drought, we too love our "sunburnt" land here in the western United States, despite its extreme and variable climate. Knowing more about how climate has changed in the American West, we can now anticipate what may be in store for the region, and we can begin preparing for that future today.

B. Lynn Ingram

ACKNOWLEDGMENTS

B. Lynn Ingram:
I would like to express gratitude to my dissertation adviser at Stanford University, James Ingle, and to Doris Sloan at the University of California, Berkeley, both of whom inspired and encouraged my research on San Francisco Bay. Roger Byrne has been a valued and supportive colleague and collaborator for almost twenty years at UC Berkeley, and he appears in many of these pages. Peter Schweikhardt, Peter Weber, Frances Malamud-Roam, Brendan Roark, Kathleen Johnson, Amy Engelbrecht, and Scott Starratt have provided inspiration and enthusiasm—in many ways teaching me much more than I have taught them over the years.

Frances Malamud-Roam:
I would like to thank my mentors and advisers, B. Lynn Ingram and Roger Byrne, and many of the other students who were so fortunate to work with Lynn and were colleagues of mine who helped me through many courses and papers. Special thanks go to Michelle Goman, Eric Edlund, Scott Starratt, and, of course, Karl Malamud-Roam.

Both authors would like to acknowledge the valued support of the late David H. Peterson at the U.S. Geological Survey, who provided inspiration and encouragement during the inception and early stages of planning and writing this book. Our special thanks go to Mike Dettinger, Michael McPhadden, Kent Lightfoot, Joan Florsheim, Jim Johnstone, Roger Byrne, Cory Phillis, Peter Weber, Malcolm Hughes, Jack Repcheck, Scott Stine,

Dan Cayan, Miriam Glickman, Don DePaolo, and Gene Ingram for valuable discussions and sage advice over the years and for reviews of portions of the draft. We are indebted to Jeffrey Mount, James Powell, and an anonymous reviewer for careful reviews of the manuscript. Alan Anderson and Paige Gilchrist provided invaluable shaping and editing of the manuscript. Many of the photographs and figures in this book were kindly provided by Arndt Schimmelmann, Jim Johnstone, Thomas Swetnam, Derrick Kelly, James Kennett, Doug Kennett, Anders Noren, Jeanne Di Leo, and Don DePaolo, for which we are grateful. Agent Amanda Menke and UC Press editor Blake Edgar provided expert guidance, advice, and valued support. Last but not least, we are extremely grateful to Dr. Wenbo Yang for helping to redraft many of the book's figures and for analyzing the isotopic compositions of many of the samples discussed in the book.

Introduction

Some say the world will end in fire,
Some say in ice.
From what I've tasted of desire
I hold with those who favor fire.
But if it had to perish twice,
I think I know enough of hate
To say that for destruction ice
Is also great
And would suffice.

ROBERT FROST, *"Fire and Ice"*

ONE OF ROBERT FROST'S MOST FAMOUS POEMS asks whether the world will end in fire or in ice. Although we are not expecting the world to end anytime soon, we anticipate a future that may be vastly different from the world we are experiencing today. We find ourselves in an interglacial period that could return to an ice age like the one that ended some 11,000 years ago; or, we could continue along a path of global warming. A climatologist might add the biblical element of flooding, which could accompany either extreme.

Whatever the future scenario, these questions have become urgent for the arid American West: where has the climate been, and where might it go from here? In *The West without Water,* we show how paleoclimatologists—scientists who study past climate—have painstakingly uncovered and deciphered clues about past weather and climate in the West to help us answer these questions.

The picture that emerges from this evidence is more complex than a choice between "fire" or "ice"—and more worrisome. With the help of dozens of paleoclimatologists whose work is described in this book, we find that even the relatively ice-free and benign "normal" weather of recent millennia has been much more variable and potentially dangerous than anything we experience today. In addition, the climate now seems to be changing rapidly toward new patterns that are certain to be more challenging for human life and dwindling ecosystems.

The terms *weather* and *climate* are often used interchangeably, but there are important distinctions. Climate refers to the statistical description of weather over a given period of time and for a given region, including the weather extremes. Weather has been monitored extensively for more than a century in the West, so we have a good idea of the region's climate for this period. For example, we know that Phoenix, Arizona, is arid, with hot summers and cool winters; San Francisco, California, has cool, foggy summers and wet, mild winters. Average high temperatures in Phoenix are much different in summer (104°F) than in winter (about 65°F), but its ten inches of rain annually are more or less evenly distributed between winter and summer storms. In contrast, San Francisco has average high temperatures that vary little throughout the year, staying mostly within the narrow range of 60–70°F. The twenty inches of rain received annually falls almost entirely between November and March, and the spring and summer months in San Francisco are often foggy.

Of course, these climatic averages cover a relatively short period of time— a mere century or so. The longer timescale climate averages have not been systematically measured by humans but have been recorded in the landscape, sediments, trees, and other natural archives, as we will discuss in later chapters.

The distinctions between weather and climate often catch people off-guard, occasionally with tragic results. Mark Twain once remarked that "climate is what we expect, weather is what we get," and the Donner Party is a famous and tragic example of "getting" weather. This group of pioneers was surprised by spectacularly abundant snow that came unusually early one year. The Illinois pioneers were headed for San Francisco in the fall of 1846 but were stopped short in the Sierra Nevada near the Nevada-California border on October 20 by a mountain range blanketed by the first snowfall of the year. Statistically, the first heavy snow storms should not have arrived so early, yet the possibility is always there. The group managed to make it as far as Truckee Lake (now called Donner Lake) where another heavy snow-storm—the second of ten major snowstorms that winter—shrouded the area, rendering these unfortunate pioneers snowbound. Another attempt to cross the summit was blocked in early November when yet another major storm increased the snowpack to ten feet. Now considered one of the most spectacular tragedies of the western migration, only 48 of the original 87 members survived, and they did so by resorting to cannibalism.

In modern parlance, the doom of the Donner Party was a result of extreme weather. Of course, during the mid-nineteenth century, just how variable and extreme the weather in the West could be was virtually unknown to pioneers entering the region. We know much more today—not just about the recent historical past but also about the longer-term climate history of the West.

Perhaps the worst climatic consequence in the twentieth century was the devastating "Dust Bowl" of the 1930s—well within living memory, yet nearly forgotten today. At its peak, this drought affected the entire western United States, from California to the Great Plains, with searing heat, roiling dust storms, and fierce winds. Along the West Coast, sparse population size, less-developed agriculture, and the buffer provided by extensive groundwater pumping meant that the effects were largely mitigated compared to the shortages and suffering such a drought would inflict today.

Looking further back, the Great Flood of 1861–62 provides an example of a climate catastrophe of biblical proportions. This flood brought quick and enduring devastation, killing thousands of people and hundreds of thousands of livestock as it submerged the entire Central Valley of California and other regions of the West under ten or more feet of water for months. But the state was relatively unpopulated at that time, and the event is virtually forgotten today.

Deadly as these recent natural assaults were, climatologists now know that they are dwarfed by ancient "megadroughts" and "megafloods" that have occurred multiple times during the Holocene, an epoch that spans 11,000 years and the entirety of recorded human civilization. Prolonged droughts, in particular—some of which lasted more than a century—brought thriving civilizations, such as the Ancestral Pueblo of the Four Corners region, to starvation, migration, and finally collapse.

Discovering and interpreting the details of recent and ancient climates is far from an easy task. The field of paleoclimatology is complex, not only because it studies conditions that vanished long ago but also because those conditions can be understood solely through multidisciplinary teamwork—including such fields as geology, chemistry, biology, hydrology, paleontology, and archaeology—using modern scientific toolkits. This book tells the story of the American West's climate detectives, scientists who work together—sharing tools, insights, and techniques—to tease out ancient clues from sediments, trees, and landscapes. This assortment of evidence allows climate scientists to piece together key features of former environments. For instance, past droughts leave evidence such as vegetation shifts, chemical clues in the sediments deposited beneath shrinking lakes, and the width and chemistry of tree rings. On the other end of the spectrum, periods of prolonged wetness cause lakes to enlarge (leaving behind ancient shorelines), floods to deposit sediment layers in floodplains, and chemical clues to be imprinted in sediments and fossils buried beneath lakes, estuaries, and the coastal ocean.

To assess how humans will need to adapt to changing conditions in the future, scientists can project what we know about past climates into the future. And what we see, through the lens of modern science, is a rapidly unfolding crisis for the West—one that demands immediate action. The rapidly accumulating evidence of past climate conditions, together with models of future conditions based on projected global warming, are overwhelming: for the American West, the times ahead are very likely to be warmer and drier—much closer to Robert Frost's fire than to his ice. The predictions also show that this generally drier climate will be punctuated by larger and more frequent floods.

We can certainly identify appropriate human responses to this emergency, many of which would require changes in how scarce supplies of water are allocated and used. We would also need to reduce our "water footprint" by becoming more aware of how much water we consume on a daily basis. But can we expect residents of the West to respond in this manner? The historical record is not encouraging. The arid West has experienced a continuing population expansion, despite early warnings by John Wesley Powell in the late nineteenth century. Powell, the one-armed geologist and explorer, led the first scientific expedition down the Colorado River in 1869 (see figure 1). After an extensive survey of the geography, climate, hydrology, vegetation, topography, and geology of the American West, Powell advised the U.S. government to abandon its ambitions of populating this arid region. In a report to the General Land Office, Powell argued that only small-scale irrigation in the West would be possible, supporting only a small number of people. Powell calculated that, even if all the water in western streams were harnessed, only a tiny fraction of western land could be irrigated.

A debate followed this report, with prosettlement boosters (especially the powerful railroad companies) arguing that the West held great promise for settlers who wanted to make a new beginning. As a rebuttal to Powell's conclusions that the West was too arid, they proposed the astonishing theory that "rain follows the plow," meaning that rainfall would increase as farmers worked the soil and planted crops, stirring up the atmosphere to produce rain-bearing clouds. The phrase, coined by amateur scientist Charles Dana

FIGURE I. The Grand Canyon, carved by the Colorado River. (Photo taken from the South Rim at Grand Canyon National Park by B. Lynn Ingram.)

Wilber, was far more popular with Congress than it was Powell. But rainfall did increase for a time in the late nineteenth century, giving the pioneers false hope. Despite Powell's admonishments, therefore, hopeful pioneers continued to flock to the West.

Many of these settlers entered the West without the benefit of knowing the long-term climate history of the region—a history we will describe in this book. Nor did they have much of an understanding of even the shorter-term weather and climate patterns. These pioneers migrated to the arid West with expectations of lush farms, easy living, and abundant water. Yet, as they came, all around them were the visible signs of an arid region prone to drought.

Conditions were particularly treacherous in the hot, dry deserts of the Southwest, catching many pioneers from the humid eastern portion of the country off-guard. Many suffered horrendous hardship on the westward journey, sustained by the hope and promise of a land of milder climate, rich farmlands, and, of course, gold. One notable group led by William Manly headed toward San Francisco during the Gold Rush in 1849. The party attempted a "short-cut" across Death Valley, south of the Sierra Nevada; clearly they had not been warned that this valley is the hottest and driest place in the Northern Hemisphere. They were lost for several weeks, surviving only by eating several of their oxen. Manly chronicled the journey, describing their constant search for water in this extremely dry region, and expressing their regret over leaving their homeland:

> We thought we could fight to the death over a water hole if we could only secure a little of the precious fluid. No one who has ever felt the extreme of thirst can imagine the distress, the despair, which it brings. I can find no words, no way to express it so others can understand.
>
> When in bed I could not keep my thoughts back from the old home I had left, where good water and a bountiful spread were always ready at the proper hour. I know I dreamed of taking a draft of cool, sweet water from a full pitcher and then woke up with my mouth and throat as dry as dust. The good home I left behind was a favorite theme about the campfire, and many a one told of the dream pictures, natural as life, that came to him of the happy Eastern home with comfort and happiness surrounding it, even if wealth was lacking. The home of the poorest man on earth was preferable to this place. Wealth was of no value here. A board of twenty dollar gold pieces could now stand before us the whole day long with no temptation to touch a single coin, for its very weight would drag us nearer death. We could purchase nothing with it and we would have cared no more for it as a thing of value than we did the desert sands. We would have given much more for some of the snow, which we could see drifting over the peak of the great snow mountains over our heads like a dusty cloud. (chap. 10)

Those great snow mountains that Manly referred to were the southern Sierra Nevada. Ironically, for these pioneers, these snow-topped mountains were ultimately the cause of their suffering. The mountain range lies from north to south, rising to an elevation of 14,000 feet, creating what is in essence a 400-mile-long barrier to the Pacific storms—the only significant source of moisture to California and much of the West. The storms move across the great Central Valley and up the slopes of the Sierra Nevada. As they climb to higher altitudes, they drop much of their precious moisture, leaving the regions to the east, particularly Death Valley (named by Manly), bone-dry. The reality can be harsh for those who live in what is called the "rain shadow," the backside of a mountain range. These can be the driest places on Earth. Manly could only look with longing at the snow drifts so unreachable and so tantalizing in their promise of abundant water.

Once the settlers of the American West had arrived, they quickly saw to the remaking of their new home, to create for themselves the promised land of plenty. In the early twentieth century, the Great Depression ushered in the great era of dam building and the expansion of major population centers across the West. The assumption behind this dramatic gesture was that water in an arid region could be "conquered and controlled" through modern engineering technology, which would prevent both droughts and floods.

Today, all of the West's great rivers are dammed, and hardly a drop of the flowing surface water remains unmanaged or uncontrolled. This new norm is insidious: the population has a collective idea of abundant water for consumption, despite what the region's forests and waterways indicate. Too few of the region's inhabitants notice the connections between dying forests, drying riverbeds, shrinking lakes and reservoirs, and the creeping drought. Water continues to flow through the taps, swimming pools are full, and vegetables are abundant and gleaming on supermarket shelves. But we should not have to wait until the reservoirs run dry to realize how tenuous our engineered control of the region's water is in the face of drought or massive floods. Engineered water management was an aid to building a large, modern society in the West during what is now understood to have been a century of benign, moderately wet times. It is most certainly not likely to sustain what is now an engorged, vastly overgrown modern society.

Drought is not the only stressor in the West, however: intense storms are also a part of its climate, giving rise to severe flooding. Growing populations and development have increased the risks of flooding in the West as urban development spreads into low-lying areas where floods have historically occurred. Some of these areas, like the floodplains of the Sacramento and San Joaquin rivers in California's Central Valley, have subsided by more than thirty feet as a result of extensive groundwater extraction, and their delta now lies below sea level. Like the Mississippi River Delta and the tragic flood in 2007 following Hurricane Katrina, these vast regions in California are increasingly susceptible to massive flooding.

Archaeological evidence shows how ancient human societies of the region coped with the vagaries of their climate over thousands of years. Communities remained small and highly mobile; they could react to flood threats by clearing out quickly. The indigenous human populations that preceded European-American settlers also avoided permanent settlements in the floodplains and deltas, using them only seasonally for fishing and hunting. During times of drought, they learned to be masters of conservation, and, if the drought was prolonged, they again could clear out to find more hospitable lands.

The fluctuations and extreme swings of climate in the past have occurred without the contributions of human activity. But the buildup of carbon dioxide and other greenhouse gases, according to most climate experts, has already begun to produce warmer and more extreme weather worldwide. Climatologists now speak in terms of even deeper droughts, and larger and more frequent floods, for the future. What we have come to expect as "normal" has not adequately prepared us as a society for what the new "normal" could present.

The message of past climates is that the range of "normal" is enormous—and we have experienced only a relatively benign portion of that range in recent history. The position of inhabitants of the West is precarious now and growing more so. To date, those who develop and enforce policy have done little in the face of all the evidence of climate research, even as the data come in with more disturbing results and urgent warnings. As we continue with an unsustainable pattern of water use, we become more vulnerable each year to a future we cannot control.

The West has experienced unprecedented rates of population growth—which was made possible by water development. Nearly all of the rivers have been dammed to control the distribution of water and to buffer the dry years. Vast reservoirs store the water that falls each winter, and extensive aqueducts transport the water throughout the region. During the early decades of the twentieth century, dry years were ignored and the water supplies were supplemented by drilling deep into buried groundwater reserves. Earthen walls along the banks of the major rivers channeled the flows, protecting settlements from the risk of floods. Through a vast network of aqueducts and canals, we have moved the water in the West to wherever we wanted it. But all of this advanced engineering cannot stop droughts from happening. Today, the Southwest has suffered under a decade of drought.

To a great extent, engineers have succeeded in their goals of taming rivers and controlling when and where the waters flow. Faucets still provide water to homes and irrigation systems, but the West continues to be dry. What are the environmental costs of maintaining the supply of water to residents? How can our knowledge of past climate variations help inform future scenarios and guide decisions for a sustainable society? We must keep in mind that the West has been plagued by droughts lasting not just years, or even decades, but more than a century. If decade- or century-long dryness revisits the region, modern society will be hard-pressed to survive without significant changes in the way we live. Now, in the twenty-first century, there is evidence that this brief time of climatic stability is slipping away, and we are entering a period of drier and more erratic conditions.

The American West remains an icon of the American frontier and is home to some of the most magnificent and bold landscapes and vast wilderness areas in the world. This book tells the story of climate in the American West: the discovery of geologic clues about past climatic changes, how they may have affected past societies, and what they may mean for the future. This history shows us that, because all human societies need water, all are ultimately dependent on climate. We have a greater chance for future prosperity if we understand our climate, its variability, and its extremes.

Part I of this book presents to the reader why climate matters to the American West, particularly when it comes to the water resources of this vast region. We begin to address the question What is the "normal" climate for the American West? In the arid to semi-arid West, water is precious but also potentially hazardous, periodically running wild with raging floods that breach natural and artificial levees, causing great suffering. When the rains fail, the rivers dry up, causing yet more suffering because of the lack of water.

Climate determines the timing, amount, and form of precipitation that a region receives; in other words, climate determines water supply. Most of what we know about climate in the West today comes from studying temperature and precipitation patterns, including extreme weather events—droughts and floods—over the past 150 years. During this period, the two worst droughts lasted six years: the heartbreaking "Dust Bowl" drought of the 1930s, and the devastating drought of 1987–92. On nature's flipside, the worst flood in the recorded history of the West occurred very early in the region's history, in 1861–62, when rainfall was three times the average annual amount and huge regions of California and the West were submerged under ten feet or more of water. What causes these extreme events?

Part II of this book tells the story of climate in the American West: the discovery of geologic clues about past changes in climate and the archaeological evidence for how those changes may have affected ancient societies. The history of climatic change over the past 20,000 years presented here highlights pioneering studies that reconstruct past precipitation, temperature, stream flows, flood events, and wildfires. Clues for these events are found in tree rings, in landforms, and in the sediments and fossils buried beneath lakes, estuaries, marshes, and the coastal ocean. In this section, we describe the methods for teasing out climate clues from these natural archives.

Evidence of change can also be gleaned from archaeological remains, such as the cliff dwellings and pueblos of the Southwest and the shell mounds along the Pacific Coast. The early people of the West lived intimately entwined with their environment, so that changes in conditions had direct consequences. Thus, when the climate archives reveal repeated prolonged droughts, the archaeological archives reveal how societies across this large region experienced and responded to environmental stresses. The greatest societal catastrophes occurred after populations expanded during years of plenty and were then struck by disastrous droughts—an unnerving parallel to modern times. On the other end of the climate spectrum is evidence of cataclysmic megafloods that struck the region every one to two centuries over the past two thousand years. In this section, we explore the possible causes for these extreme events and the climate cycles revealed by the natural archives.

Whereas Parts I and II set the stage for understanding climate in the West and how it has affected societies in the past, Part III turns to today's growing water crisis. The roots of this crisis lie in the westward expansion of the United States and the exponential growth of the region's population. Humans have moved into the deserts, floodplains, and deltas of the West, exacting a high cost to the natural environment, particularly aquatic ecosystems. We describe how a warmer future may affect the West's water resources. Less snow in the mountains, earlier snowmelt in the spring, drier summers, increased forest fires, and shifted storm tracks are among the changes predicted. Finally, we consider what lessons can be learned from the climate history that we have constructed in our research, and we discuss potential solutions to the current and future water crises.

PART ONE

———

Floods and Droughts in Living Memory

Climate matters. It mattered to the Vikings, who settled Iceland, explored the New World, and were lured north to Greenland during a period of unusually warm weather a millennium ago. Climate mattered to Oklahoma farmers during the Dust Bowl years of the 1930s, when many people headed west as much of their soils headed east on the withering winds. Today, with floods and drought, feast and famine, climate matters to many of us much of the time.

RICHARD ALLEY, *The Two-Mile Time Machine*

ONE

From Drought to Deluge

"NORMAL" CLIMATE IN THE WEST

And it never failed that during the dry years the people
forgot about the rich years, and during the wet years they lost all
memory of the dry years. It was always that way.

JOHN STEINBECK, *East of Eden*

HUMANS TEND TO PERCEIVE CLIMATE AS A FORCE of nature, one
to be measured, classified, and ultimately conquered. John Steinbeck's quote
speaks to another human characteristic: the tendency to forget. This seems
especially true when it comes to the periods of wet and dry over the past
century or so in the American West.

Perhaps, however, we can be sympathetic to this tendency when we
realize that to "measure the climate" at any given point is really a measurement of the current *weather*. To understand complex climatic patterns requires a much longer timescale. Thus, in our lifetimes, we experience only a small sampling of climate effects—not the full range—in
the West. Imagine a performance of a Beethoven symphony with
only the first chords, played over and over again. If that were all we heard
of the symphony, we would believe it was beautiful—but simple. We could
hum along and predict what would come after the initial chord sequence.
However, as we know, Beethoven wrote complex and intricate symphonies, often full of surprises. These may be pleasurable in musical compositions, but surprises in climate can have catastrophic results for societies.
We can remove the element of surprise by understanding as much about
our climate as possible.

In this book, we focus primarily on those aspects of climate that have
an impact on water resources in the West. To further our understanding of
those resources, imagine "flying over" North America using Google Earth.
Traveling from east to west, we see obvious contrasts: shades of green that
dominate the eastern portion of the country abruptly give way to vast areas

FIGURE 2. Donner Lake in northeastern California, the eastern Sierra Nevada, eleven miles northwest of Lake Tahoe. (Photo by B. Lynn Ingram.)

of brown, tan, and yellow. The color contrasts are due, of course, to moisture differences. The humid East supports more forests and, moving westward, wide prairies. Western North America is more constrained by water availability, especially in the southwestern deserts.

If we zoom in a bit closer and fly across this western region from west to east, we begin to see a relationship between the lay of the land and the colors of the land—a relationship that is largely based on water. Passing above the Pacific coast, we observe mountains, especially along the northern coastline, with lush rainforests that give way quickly to grasslands and then to deserts. Within this region, it is clear where the water falls: the Pacific Northwest and Northern California appear deep green, as does the central and southern Sierra Nevada mountain range (see figure 2). Much of Southern California and the states of Nevada, Utah, Arizona, and New Mexico appear in varying shades of brown, tan, and red, reflecting the vegetation and the composition of the soils—both related in large part to the lower amounts of rain and snow that fall there each year (see figure 3).

Precipitation over the American West is spatially irregular because it depends on the complex behavior of the Pacific Ocean and its exchange of

FIGURE 3. Monument Valley, southern Utah. (Photo by B. Lynn Ingram.)

water and energy with the atmosphere. The atmosphere receives moisture from the ocean through the process of evaporation, and some of that moisture is eventually carried away from the ocean and delivered on the continent as rain or snow. Just how and when that evaporated seawater drops as rain or snow over the West depends on the patterns of ocean currents and atmospheric winds and on the interaction of moisture-laden air with the landscape as it advances inland.

Weather satellite images of the West Coast vividly show the movement of the atmosphere over the Pacific Ocean: large, counterclockwise swirls of white, moisture-laden air that drift toward the coast and finally over the land. The topography of this region, especially the high mountain ranges, disrupts this flow of air, forcing it to rise up the slopes. As the air moves to higher altitudes, the moisture condenses, because rising cooling air cannot hold as much water vapor. The major mountain chains of the West—the Cascades in the northwest, the Sierra Nevada in California, and the Rockies farther east—are oriented north-to-south or northwest-to-southeast, which are the ideal orientations for intercepting the Pacific winter storms that flow from west to east. As these storms are forced up and over these mountain chains, much of the moisture is wrung out on

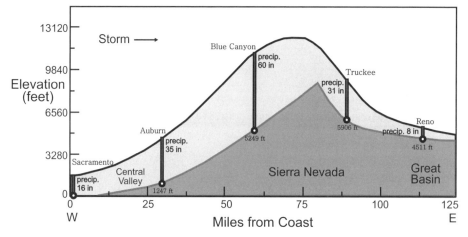

FIGURE 4. Schematic west-to-east cross-section from the California coast to the Nevada border, showing the elevation of the Sierra Nevada and corresponding annual average levels of precipitation across the range. (Sketch by B. Lynn Ingram.)

the windward slopes, leaving the leeward slopes and places to the east relatively dry (see figure 4). This explains the relatively lush western slopes of the mountains in the West compared to the drier and sparsely vegetated eastern slopes and valleys beyond (see figure 5). In this way, the landscapes of the American West both have an influence on the region's weather and are formed by it.

California spans nearly ten degrees of latitude, which results in a wide spectrum of natural plant communities, determined largely by the rainfall regime. Storm systems entering California first encounter the Coast Ranges and then pass through the Central Valley, move up the high mountain ranges of the Sierra Nevada, and continue into the deserts of the Great Basin. Northern California is the southern extension of the lush temperate rainforests of the Pacific Northwest, whereas Southern California is part of the deserts of the semiarid to arid Southwest. Nevertheless, California's moisture regime follows strict—if somewhat unusual—patterns. The Department of Geography at the University of California, Berkeley, has a long history of researchers interested in climate, including Jim Parsons, who, in a personal communication with this book's authors, noted that the state's climate is unique because of its arbitrarily conceived boundaries, which outline the only area of winter rain and summer drought in North America. This so-called Mediterranean climate—warm, dry summers and mild, wet winters—is a common occur-

FIGURE 5. The Owens Valley east of the Sierra Nevada (with eastern Sierra Nevada peaks in the background). (Photo by D. J. DePaolo, University of California, Berkeley.)

rence on the west coasts of continents at about 35 degrees latitude. During the summer months, the atmosphere over California is dominated by the Pacific High, an area of warm, dry air that prevents storms from entering the state. Thus, California's climate is characterized by two main seasons: one wet, the other dry.

As described above, Pacific storms in the winter drop much of their rain and snow along the western mountain slopes, concentrated in a relatively short wet season. In fact, California has the highest year-to-year variability in precipitation of all the western states. This variability is the result of a relatively small number of Pacific storms—on average five or six—that enter the state between the late fall and early spring. California's precarious water situation is apparent in the fact that a shortfall of just one or two major storms can mean the difference between a "normal" water year and a dry year. As a result, water management is particularly challenging in this state.

In addition to the year-to-year variations in precipitation, of course, California's climate changes on longer timescales, from decade to decade, century to century, and millennium to millennium. The patterns of these longer-term variations are discussed in more detail later in this book.

In the higher mountain elevations of the West, precipitation often falls as snow. This snow provides a crucial amount of water, filling streams, lakes, and reservoirs to be tapped during the long, dry summer. Across the West, the melting of the winter snowpack provides a significant proportion—50 to 80 percent—of the annual water supply for ecosystems as well as human consumption.

Water managers throughout the West rely on early forecasts of snowpack to plan for each water year. These forecasts are based on measurements of the mountain snowpack, obtained by hardy snow surveyors who head out on skis into the mountains to determine how much meltwater will be available in the coming year. Each year in California, over 300 snow courses are checked, each 1,000 or more feet long with measurements made every 100 feet on average. The practice has been a part of the state's water management program since the early 1900s, and the data collected provide valuable planning information for farmers, utility companies, municipal water managers, and flood control managers.

During the 1950s, 1960s, and early 1970s, another researcher in the UC Berkeley Department of Geography, Douglas Powell, spent his winters on skis taking part in the annual state snow surveys. Clear, sunny fall days may be welcome to most Californians, but Powell was always happier when the storm clouds arrived early, heralding a wet year that would replenish the state's water supplies. He would often remind his summer-session students, with a hint of wistful optimism, that storms—even thunderstorms—could arrive as early as September. Moreover, autumn blizzards in the Sierra Nevada range are not unheard of: in 1846, a series of October blizzards dropped ten feet of snow, trapping the Donner Party during their ill-fated trek across the Sierra.

Few of Powell's colleagues or students (this book's authors among them) have experienced the linkages between landscape and climate as intimately as he did. For several weeks each year, he and his team strapped on cross-country skis and backpacks loaded with log and field notebooks, pens, and special snow collection tubes to determine the depth, water content, and density of the snow he would trek across each day. They would measure the water content of the snow by filling a special tube with snow, weighing it, and weighing the empty tube. One ounce of snow translates into one inch of water content of the snow, so the snowpack can be converted into the amount

FIGURE 6. Snow surveyors measuring snow depth and water content in the Sierra Nevada, California. (Photo by Doug Powell, courtesy of the Department of Geography photo archives at the University of California, Berkeley.)

of water that will be generated when that snow melts (see figure 6). Powell could think of no better way to spend his winter months than in the high Sierra Nevada, measuring the precious gift that the mountains held in reserve for the long, dry summer and fall months of the West.

Snow surveyors are still important today, but, since 1977, a considerable amount of data is collected by a network of instruments that have been installed across the mountains of the West to forecast water supply. These instruments automatically transmit the snow and climate information they measure to a centralized computer system by radio telemetry. Changes in daily snowpack are monitored, along with temperature, precipitation, stream flow, reservoir and groundwater levels, and soil moisture—all part of an early warning system for floods and droughts in the western United States.

As we will discuss further in chapter 13, studies of changes in the snowpack over the past century reveal that the average snowpack has steadily decreased by about 10 percent over that period, most likely due to warmer temperatures. Climate modeling predicts that, by the end of the twenty-first century, the snowpack could shrink by another 40 to 80 percent.

Scientists began closely monitoring the West's climate after the nineteenth century, recording the region's weather variables: temperature, atmospheric pressure, precipitation, and wind speed and direction. These measurements provided monthly and yearly averages, defining the region's climate. Climatologists have also examined extreme deviations from the mean in order to understand the causes of these deviations and perhaps learn how to predict them. The measuring and monitoring of climate over the past century has allowed climate scientists to assemble the pieces into a larger picture that begins to explain what drives climate, including the critical interactions between the Pacific Ocean and its overlying atmosphere. As geographically varied as the West is, climate in this region has been fairly consistent over the past 150 years and, by and large, suitable for human habitation. This period of time would seem to be long enough for us to know what to expect from our climate, both in normal years and in "extreme" years. However, as we shall see later in the book, this is not the case.

Extreme events have struck every few decades over the past century and a half, some with catastrophic results. Even more extreme events have occurred with some regularity in the distant past, though the intervals between them often lasted a century or more. Understanding the causes of these events and how they affected the West in the past can help current residents anticipate the effects when such events recur in the future—which they surely will.

Droughts

On one end of the climate spectrum is drought, when the wet season fails partially or altogether. Nothing brings more fear to the hearts of water planners in the American West than drought. Droughts lack the visual and visceral images of other natural disasters; instead, they creep up on the landscape with no clear beginning or end. Meteorologists define drought rather vaguely as an abnormally long period of insufficient rainfall that adversely affects growing or living conditions. Such a bland definition belies the devastation wrought by these unique natural disasters. In human history, whole civilizations, even while at their peak, have toppled in the face of prolonged drought. For humans, drought means failed crops, desiccated landscapes, water rationing, decimated livestock, and, in severe cases, water wars, famine, and mass migration.

For the more distant past, drought is defined in relative terms, such as when precipitation, runoff, and lake levels fall below the long-term average by at least one standard deviation, or when the growth rings of trees become narrower than average by a certain amount. In drought-prone regions, native plants and animals usually employ adaptations to survive through periods of lower than average precipitation; such adaptations have evolved over very long periods of time. Humans in the West today, in contrast, have flourished only by manipulating the natural hydrology using modern engineering and technology.

The Driest Single Year in the West: 1976–1977

The worst single year of drought in recorded history over much of the American West occurred during the winter of 1976–77. The drought was widespread, extending across the entire state of California and to the midwestern states and north into the Canadian prairie region north of Montana. Even the Pacific Northwest experienced its worst drought on record, leading to shortfalls in crop yields and region-wide water rationing. Record low flows were measured on the Columbia River, leading to restrictions on power consumption from hydroelectric plants. The snowpack in mountains from California to Colorado reached historic lows, and ski resorts from the Sierra Nevada to the Rockies remained closed for much of the season.

The 1976–77 drought was particularly devastating in California. Precipitation levels that year dropped to less than half of the average level throughout the state (see figure 7). Northern California was especially hard-hit, with rainfall reaching only 15 percent of the average in some regions. Runoff and river flow in California fell to a quarter of their normal levels, and storage reservoirs saw dramatic declines, falling to about a third of their normal levels. In the Sierra Nevada, Lake Tahoe dropped below its outlet at the Truckee River (near Tahoe City), causing the Truckee River to cease flowing, which in turn caused Pyramid Lake in northwestern Nevada to drop as well. Aquatic habitats were severely affected, with the worst impacts in the lower to mid-elevations of the Coast Ranges and the western slope of the Sierra Nevada. Low river flows, combined with higher water temperatures, severely disrupted the migration and spawning of migratory fish species, including all species of salmon and trout.

The shortage of surface water in the state prompted an increased use of groundwater for agriculture and cities. The proportion of groundwater used

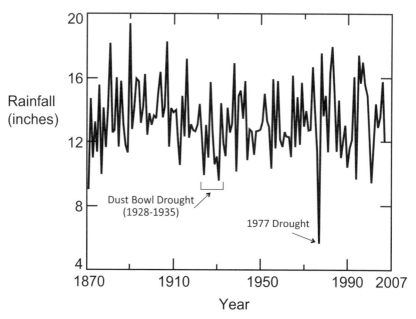

FIGURE 7. Precipitation during winter (November through April) from 373 stations in the western United States from 1871 to 2007. The year 1976–77 was the driest on record (including paleoclimate records from the past 500 years). (Graph courtesy of Dr. James Johnstone at the Joint Institute for the Study of the Atmosphere and Ocean, University of Washington.)

increased from 40 percent of the total amount of water consumed during a normal year to 53 percent, causing a precipitous drop in the water table throughout California. Particularly hard-hit was the southern Central Valley (the San Joaquin Valley), where the water table dropped so low (by up to fifty feet) that pumping capability could not keep up with demand from irrigated agriculture, and groundwater depletion was estimated to be five million acre-feet, or four times the annual average. Nine thousand water wells were deepened or drilled. In order to aid in water efficiency, many farmers shifted from growing rice and other water-intensive crops to grains, particularly barley, wheat, and oats.

Well over seven and a half million trees died as a result of the 1976–77 drought. Some regions in the highest elevations of the Sierra Nevada lost three-fourths of their trees, including ponderosa and Jeffrey pines, red firs, and white firs. These trees, weakened by the lack of water, succumbed to insect-related diseases like bark beetle or to longer-term conditions such as mistletoe and root rot. Wildfires also claimed many trees. In August 1977 alone, lightning strikes sparked 1,400 wildfires in Northern and Central California forests.

In Central California, the dry year meant record low freshwater flows to the Sacramento–San Joaquin Delta and San Francisco Bay, with cascading impacts on aquatic ecosystems. Water quality declined throughout the delta, where water is pumped and transported to Southern California, providing drinking water for twenty million Californians. With lower river flows, seawater penetrated upstream far into the delta, causing chloride concentrations to quadruple. In an effort to prevent this, water engineers released as much freshwater from the meager reservoir supplies upstream as possible and constructed temporary barriers in the delta. Nevertheless, severely reduced freshwater inflow raised the water salinity in the San Francisco Bay to record high levels, leading to a dramatic decline in the bay's phytoplankton, the base of the estuarine food chain. Fish populations in the estuary crashed as their primary food sources disappeared. In both the bay and the delta, decreased freshwater flows also increased the concentrations of pollutants, including pesticides drained from farm irrigation fields in the Central Valley. The effects of that extreme year spread like tentacles of pain deep into the ecosystems of the West.

California's economy also suffered that year: in the Sierra Nevada, the sparse snowfall had severe impacts on the business of winter sports and the tourism industry. Ski resorts in Northern and Central California were particularly hard-hit. Some of these resorts never opened, and those that did were forced to close within a very short period. The summer recreation economy also suffered, since resorts on the state's rivers and reservoirs were affected by low water levels and forest closures. Many of those resorts closed, some declaring bankruptcy, as shorelines receded, rivers dried up, and boat launches and docks were left high and dry. Fishing, rafting, and other river activities were not possible as flows diminished to a trickle during the summer and fall of 1977.

Forty-seven of California's fifty-eight counties declared a state of emergency, and severe water restrictions and mandatory rationing were imposed. In Southern California, the Metropolitan Water District was forced to increase imports from the Colorado River to near capacity. That summer, the Colorado River aqueduct supplied 85 percent of the water delivered to Los Angeles, Orange, Riverside, and San Diego counties.

So desperate was the situation that the state considered increasingly far-fetched and elaborate schemes for getting more water, including importing water by pipeline or railroad from as far away as Alaska's Yukon River, or towing icebergs to the region from Antarctica. Cloud-seeding was tried in Northern California. Losses to California's economy eventually reached $2.6 billion.

But there was a silver lining in all of this gloom. Due to the enormous economic costs, California and other states in the West became serious—for the first time—about water conservation. Low-flush toilets and other water conservation measures got their start. Industries realized that implementing water conservation measures would be the most effective way to meet future water shortages. For instance, the agriculture industry, which consumes 80 percent of California's freshwater resources, began implementing soil and water conservation practices.

Floods

On the other end of the climate spectrum, an overabundance of precipitation and runoff in the West can also prove disastrous, causing massive flooding. Floods are common in the West; seldom has a decade passed free of these devastating events. During unusually wet years, rivers overtop their channels, inundating vast reaches of land that are typically dry and increasingly populated. The enormous power of a flood destroys everything in its path, leading to landslides and mudflows, washed-out roads and bridges, lost crops, lost homes, even lost lives. Worldwide, floods cause more deaths each year than any other natural disaster.

Climatologists talk about droughts and floods as two sides of the same coin, because the worst floods in the West often followed on the heels of some of the most severe and prolonged droughts. In 1937–38, for instance, catastrophic floods hit California and other parts of the West, punctuating the end of the infamous Dust Bowl drought. The northern two-thirds of California suffered severe flooding in December 1937, with record flows on rivers draining the northern and central Sierra Nevada. Two strong Pacific storms dumped ten inches of rain in Southern California the following February and March, resulting in catastrophic flooding from San Diego to Los Angeles and into the Mojave Desert. Those floods killed 110 people and destroyed almost 6,000 homes.

Next came severe mudslides, destroying all roads leading to the San Gabriel Mountains, stranding residents for days, including an elementary school class forced to climb on top of the cabinets in their classroom to get away from the rising muddy waters that covered their desks. Following a sleepless night listening to the roar of a small river, the children went outside in the morning to a scene of devastation: entire houses and the remains of houses were strewn haphazardly across the landscape. Farther downstream, the city of Anaheim lay beneath five feet of water, and residents had no choice but to spend the night on their rooftops. The unusually wet conditions of that winter of 1938 extended into southern Arizona, where precipitation was 74 percent above normal and the Gila, Salt, and Verde rivers all flooded their banks.

This scenario was repeated two decades later when a period of drought in the 1950s was followed by massive floods in 1955. The central Sierra Nevada and southern San Francisco Bay regions of California were particularly hard-hit. Record rainfall in December of that year fell across the state, causing massive flows on Central Valley rivers. In Southern California, Orange County's biggest natural disaster occurred as warm rain melted snow in the San Bernardino Mountains, causing catastrophic flooding along the Santa Ana River. Once again a state of disaster was declared for all of California.

Devastating floods are an inevitable characteristic of the climate in the West, a topic we will discuss in future chapters. Despite a century of elaborate engineering that has attempted to tame and control the flow of water in the region, floods continue to inundate the western United States with a frequency that seems to have increased. Yet the floods that have been witnessed and recorded in the West over the past 100 or so years represent a very small sampling of the deluges and floods that are an integral part of the region's hydrology. Climate researchers are finding increasing evidence for much larger events that have occurred with alarming regularity over the past several millennia.

It must be borne in mind, however, that, despite their destructive potential, floods are a natural part of both the hydrologic and the ecological systems of the West. Floodwaters carry nutrients and organic material that are deposited onto the surrounding floodplains, producing fertile soils. These soils have made California one of the most important agricultural centers in the world, generating $30 billion per year. More than half of the fruits, nuts, and vegetables consumed in the United States are grown on some 87,500 farms in California's Central Valley.

Nevertheless, residents in California and other regions of the American West continue to face risks from catastrophic flooding as they place themselves directly in the path of floodwaters, building their homes on floodplains that will inevitably be washed away in the next extreme wet year. To this day, cities sprawl onto the floodplains, housing developments pop up in the deltas, and homes are built on the edges of cliffs and canyons where landslides occur.

Clearly, as Steinbeck indicated, society has collectively "lost its memory" of the earth's climatic past. The worst flood in recorded history in the West occurred in the winter of 1861–62. This diluvial disaster turned enormous regions of California into inland seas for months. Today, these same regions are home to California's fastest-growing cities. Much can still be learned from that historic flood, since many of the survivors wrote chilling accounts of it. In the next chapter, we focus on the flood of 1861–62, relating firsthand accounts of harrowing events and drawing the lessons we should have learned from this catastrophe.

TWO

The 1861–1862 Floods

LESSONS LOST

> The great central valley of the state is under water—the Sacramento and San Joaquin valleys—a region 250 to 300 miles long and an average of at least twenty miles wide, or probably three to three and a half millions of acres!
>
> WILLIAM BREWER, *Up and Down California in 1860–1864*

THE SOUTHERN CALIFORNIA CATACLYSM

THE ROAR OF THE WATER far up the valley woke Father Borgatta during the night. He ran through the driving rain up the hill next to his little church and frantically began to ring the church bell to warn the sleeping village. The villagers—men, women, and children—fled the rising waters on foot, but the tempestuous floodwaters, billowing fifty feet high, crashed through the village, almost catching the fleeing residents attempting to run to safety.

The rains had started innocently enough on Christmas Eve, 1861. But they continued day and night for the next twenty days, culminating in a weeklong downpour that swept this Southern Californian village off the map. The village was located along the Santa Fe Trail between New Mexico and Los Angeles, about five miles northeast of present-day Riverside. Founded in 1845, Agua Mansa ("calm water," in Spanish) was situated in a peaceful river valley surrounded by lush riparian vegetation: sycamore, willow, and cottonwood trees graced the village, along with the vineyards and orchards planted by the villagers. The flood obliterated all of that in one tragic hour: the traditional adobe houses seemed to dissolve into the brown, churning water; large trees were uprooted and carried away, along with horses, cows, and sheep.

All that remained of this idyllic village were the front steps and two marble pillars of Father Borgatta's church, bearing testimony to the highest elevation the Santa Ana River reached on that devastating night. Almost a century later, this high-water mark allowed hydrologists, commissioned by San Bernardino County, to calculate the peak flow of this flood: an incredible 330,000 cubic feet per second, the highest ever recorded for the Santa Ana River. The torrent swept away settlements in San Bernardino, Riverside, and Orange counties, and it submerged vast areas that lay downriver.

Ironically, farmers and ranchers had prayed for rain in the years preceding the great floods of 1861–62, because almost two decades leading up to that year had been exceptionally dry. But in December 1861, the farmers' prayers were answered with a vengeance. A series of monstrous Pacific storms slammed, one after another, into the west coast of North America, beginning in Canada in November and working its way as far south as Mexico, producing the most violent flooding residents ever saw—before or since. Sixty-six inches of rain fell in Los Angeles that year, more than four times the normal amount. Thirty-five inches fell in the 30 days between December 24, 1861, and January 23, 1862.

All across Southern California that wet winter, rivers surged over their banks, spreading muddy water for miles across the arid landscape. Large brown lakes formed on the normally dry plains between Los Angeles and the Pacific Ocean, even covering vast areas of the Mojave Desert. In Anaheim, Orange County, flooding of the Santa Ana River created an inland sea four feet deep stretching up to four miles from the river and lasting four weeks. In San Diego, just north of the Mexican border, rainfall was three times the normal amount for the month of January, causing rivers to spill across the surrounding countryside, washing out roads, drowning cattle, and isolating people when travelling by wagon became impossible. All of Mission Valley was under water, and Old Town was evacuated as rising waters filled the town.

NORTHERN CALIFORNIA UNDER WATER

Given how quickly news spreads today in our digital world, it seems incredible that news of Southern California's floods that awful winter did not reach Northern California until February 1, 1862. But residents there, where most of California's 500,000 people lived, were contending with devastation and suffering of their own.

Before the floods, California state geologist Josiah Whitney had hired an assistant, William Brewer, to help survey the young state's natural resources.

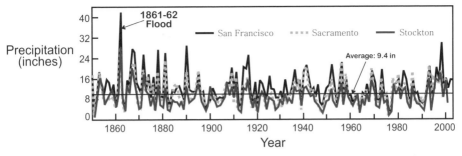

FIGURE 8. Precipitation in San Francisco, Sacramento, and Stockton, California, from 1850 to 2003. Precipitation during the winter of 1861–62 was three times higher than the historical average. (Redrawn by B. Lynn Ingram based on Mock 2006.)

In 1862, Brewer sent a series of letters to his brother on the East Coast describing the surreal scenes of tragedy that he witnessed during his travels in California that winter and spring. Brewer's letters documented the unprecedented snowfall in November and early December 1861 that blanketed the Sierra Nevada range. The snow did not last long, however, because the same series of warm storms that wreaked havoc along the West Coast also melted the snow and drenched the western slopes of the Sierra Nevada with 60–102 inches of rain. Nevada City, fifty miles northeast of Sacramento, received a total of more than nine feet of rain for the season (whereas the normal rainfall there is fifty-five inches, or just under five feet).

Brewer commented on his amazement that the two-month rainfall totals in California that year were more than *two years* of rainfall in his much soggier hometown of Ithaca, New York. San Francisco, Sacramento, and Stockton received almost four times their average rainfall for the months of December to February (see figure 8).

The series of warm storms swelled the rivers in the Sierra Nevada range so that they became raging torrents, sweeping away entire communities and mining settlements in the foothills of the Sierra Nevada—California's famous "Gold Country." A January 15, 1862, report from the *Nelson Point Correspondence* described the scene: "On Friday last, we were visited by the most destructive and devastating flood that has ever been the lot of 'white' men to see in this part of the country. Feather River reached the height of 9 feet more than was ever known by the 'oldest inhabitant,' carrying away bridges, camps, stores, saloon, restaurant, and much real-estate." Drowning deaths occurred every day on the Feather, Yuba, and American rivers. In one tragic account, an entire settlement of Chinese miners was drowned by floods on the Yuba River.

FIGURE 9. Map of California showing the approximate areas of flooding in 1861–62. The white regions on the map show the lowest elevations in the state, most of which were under water during the flood. (Map redrawn by B. Lynn Ingram.)

The Central Valley Submerged

In the Sierra Nevada, cold arctic storms dumped ten to fifteen feet of snow that winter, and these were soon followed by atmospheric river storms, which, as will be discussed later in this chapter, poured heavy warm rains on the snow. Unable to penetrate the still-frozen soils, an enormous pulse of water from the rain and melted snow flowed downslope and across the landscape, overwhelming streams and rivers and creating a huge inland sea in California's enormous Central Valley—a region at least 400 miles long and up to 70 miles wide (see figure 9). This vast, temporary inland sea covered farmlands and towns, drowning people, horses, and cattle, and washing away houses, buildings, barns, fences, and bridges. On January 18, 1862, the following appeared in the *Contra Costa Gazette*:

The havoc of the flood has been sadly general. Along the river courses, where boats are common, a ready means of escape and safety has been at hand. Not so, however, in the interior. Witness the extensive plains of the Sacramento valley, for example, where for miles and miles in every direction no highlands are to be seen, and where fences, barns, hay and grain as well as farm houses, cattle, hogs and horses have been all inundated and often carried hopelessly away. Very many lives have been lost and an immense amount of property utterly ruined and destroyed. Certainly not less than one-third of the surface of California has been visited as if by a blighting plague, a desolation earthquake or a devouring fire. (p. 2)

The water was so high it completely submerged telegraph poles that had just been installed between San Francisco and New York, causing a complete break in transportation and communications over much of the state for a month. The entire Central Valley was transformed into a lake, submerging thousands of farms, drowning cattle, and rendering roads impassable. On February 9, 1862, Brewer wrote:

Nearly every house and farm over this immense region is gone. There was such a body of water—250 to 300 miles long and 20 to 60 miles wide, the water ice cold and muddy—that the winds made high waves which beat the farm homes in pieces. America has never before seen such desolation by flood as this has been, and seldom has the Old World seen the like. (p. 244)

A great sheet of brown, rippling water extended from the Coast Range to the Sierra Nevada. One-quarter of the state's estimated 800,000 cattle drowned in the flood, marking the beginning of the end of the cattle-based ranchero society in California. Drought the following year finished it off. One-third of the state's property was destroyed; seven homes out of every eight were severely damaged; and one home in eight was destroyed completely or carried away by the floodwaters.

Sacramento under Water

In 1861, the nation was in the first year of a civil war that would consume the attention of the northern and southern states. Out West, California was a fledgling state, only in its twelfth year. The Gold Rush that had drawn so many people to the region had been going on for thirteen years. Sacramento, a hundred miles northeast of San Francisco and located at the confluence of the Sacramento and American rivers, was in many ways a hub: the young state's sparkling new capital; an important commercial and agricultural center; and the terminus for stagecoaches, wagon trains, the pony express, and riverboats from San Francisco. The newly installed telegraph connecting California with New York was located in Sacramento.

Of course, floods in Sacramento were not unknown to the residents, yet nothing could have prepared them for the deluges and massive flooding that engulfed the city that winter of 1861–62. Alfred Doten of Milpitas described the scene in his diary:

> After the inundations of the memorable winter of '52–3, a high levee or embankment was built around the city, in order to protect it from all future floods. It answered very well for such fine winters as we have had since then, but this winter it could not fence out the big waters. The raging floods of the American, and Sacramento, attacked the devoted city, both in the rear and in the front. Over the levee came the leaping waters, scornfully laughing at the puny obstructions that the presumptuous hand of man had placed in its way. The alarm bells sounded, and people rushed hither and thither to save what they could, or to note the progress.
>
> Along the street levels, and over the sidewalks, rushed the gliding demon of destruction, submerging street after street, until at last it had complete possession of the doomed city. Still and steadily rose the water over the curbstones, over the doorsteps, and into the houses, stores, and hotels it rushed. Many small wooden buildings floated off down stream, and some of the walls of the brick buildings, becoming gradually undermined by the action of the current, settled and fell. Boats were out everywhere, rescuing those who were in danger. (Van Tilburg Clark 1973, 648)

The floodwaters rose one foot per hour throughout the afternoon, eventually submerging the entire city of Sacramento under ten feet of brown, debris-laden water. The water was so deep and dirty that no one dared to move about the city except by boat (see figure 10). Indeed, the entire Sacramento Valley was submerged from the foothills of the Sierra Nevada to the Coast Ranges, with only some hills and Native American shell mounds peaking up from the water like islands.

California's new governor, Leland Stanford, was to be inaugurated on January 10, 1862, but floodwaters swept through Sacramento on the day of his inauguration, submerging the city. As citizens fled the city by any means possible, the inauguration ceremony took place at the capital building despite

FIGURE 10. Flood of 1861–62 in Sacramento, California. Lithograph of K Street in Sacramento, near Front Street. (Lithograph courtesy of Bancroft Library Archives, University of California, Berkeley.)

the mounting catastrophe. Governor Stanford traveled from his mansion to the capital building by rowboat. Following the expedited ceremony, with floodwaters rising all around at the rate of one foot per hour, Stanford rowed back to his mansion, where he was forced to steer his boat to a second-story window in order to enter his home.

William Brewer was staying in San Francisco at the time and described Sacramento's plight in a letter to his brother on January 19, 1862:

> The rains continue, and since I last wrote the floods have been far worse than before. Sacramento and many other towns and cities have again been overflowed, and after the waters had abated somewhat they are again up. That doomed city is in all probability again under water today.
>
> Sacramento is out of food. Benevolent societies are active, boats have been sent up, and thousands are fleeing to this city. There have been some of the most stupendous charities I have ever seen. An example will suffice. A week ago today news came down by steamer of a worse condition at Sacramento than was anticipated. The news came at nine o'clock at night. Men went to work, and before daylight tons of provisions were ready—eleven thousand

pounds of ham were cooked. Before night two steamers, with over thirty tons of cooked and prepared provisions, twenty-two tons of clothing, several thousand dollars in money, and boats with crews, etc., were under way for the devastated city. (pp. 241–42)

Conditions did not improve in the following weeks. California's legislature, unable to function in the submerged city, finally gave up and moved to San Francisco on January 22 to wait out the floods.

Sacramento remained under water for six months. Brewer visited the city on March 9, 1862, and described the scene:

> Such a desolate scene I hope to never see again. Most of the city is still under water, and has been there for three months. A part is out of the water, that is, the streets are above water, but every low place is full—cellars and yards are full, houses and walls wet, everything uncomfortable. No description that I can write will give you any adequate conception of the discomfort and wretchedness this must give rise to. I took a boat and two boys, and we rowed about for an hour or two. Houses, stores, stables, everything, were surrounded by water. Yards were ponds enclosed by dilapidated, muddy, slimy fences; household furniture, chairs, tables, sofas, the fragments of houses, were floating in the muddy waters or lodged in nooks and corners—I saw three sofas floating in different yards. The basements of the better class of houses were half full of water, and through the windows, one could see chairs, tables, bedsteads, etc., afloat. Through the windows of a schoolhouse I saw the benches and desks afloat. Over most of the city boats are still the only way of getting around.
>
> It is with the poorer classes that this is the worst. Many of the one-story houses are entirely uninhabitable; others, where the floors are above the water are, at best, most wretched places in which to live. The new Capital is far out in the water—the Governor's house stands as in a lake—churches, public buildings, private buildings, everything, are wet or in the water. Not a road leading from the city is passable, business is at a dead standstill, everything looks forlorn and wretched. Many houses have partially toppled over; some have been carried from their foundations, several streets (now avenues of water) are blocked up with houses that have floated in them, dead animals lie about here and there—a dreadful picture. I don't think the city will ever rise from the shock, I don't see how it can. (p. 249)

Charitable organizations, such as the Saint Mary's Sisters of San Francisco, scouted the disaster areas by boat to aid the flood victims, delivering the injured and homeless to their hospital in San Francisco. They also provided refuge for the many orphans whose parents were killed in the deadly floods.

The death and destruction of this flood caused such trauma that the city of Sacramento embarked on a long-term project of raising the downtown district by ten to fifteen feet in the seven years after the flood. Governor Stanford also raised his mansion from two to three stories—leaving empty the ground floor, to avoid damage from any future flooding events.

The San Francisco Bay Area Deluges

Downstream of Sacramento, towns and villages throughout the San Francisco Bay Area were struggling with catastrophes of their own. In Moraga, a small community twenty miles east of the San Francisco Bay, great flooding and destruction washed away meadows and drowned horses and cattle.

The scene was repeated throughout the greater Bay Area: twenty miles northeast of San Francisco, four feet of water covered the entire town of Napa; to the east, the small town of Rio Vista on the Sacramento River was under six feet of water. The entire population of Alamo, at the foot of Mt. Diablo fifty miles east of San Francisco, was forced to flee rising floodwaters. People abandoned their homes in the middle of the night; some found refuge, others drowned. The San Ramon Valley was one sheet of water from hill to hill as far as the eye could see. The destructive force of the floods carried houses, otherwise intact and complete with their contents, away in the rapids, and horses, cattle, and barns were swept downstream for miles.

The heavy rains also triggered landslides and mudslides on California's steep hillsides. For instance, in Knights Ferry and Mokelumne Hill nearly every building was torn from its foundation and carried off by mud and rock, and a major landslide occurred at the town of Volcano in the Sierra foothills, killing seven people.

The Gold Rush Connection

In an ironic twist of fate, the California Gold Rush, which had drawn so many people to California and heralded modern society in the West, was partly to blame for the extensive damage brought on by the floods in Northern California that winter of 1861–62. Early mining practices used a simple process of panning for gold, but once the accessible gold was cleared from streams draining the western foothills of the Sierra Nevada, the miners turned to more intensive techniques to increase their yield. The invention of "hydraulic gold mining" allowed mining industries to get at the gold still in the mountain rocks. Giant hoses blasted the western slopes of the Sierra Nevada, eroding hillsides and generating large amounts of sediment.

Log dams were built to catch the debris in the streams before it reached major rivers. During the fateful winter of 1861–62, the log dams failed, releasing a huge pulse of sediment—boulders, cobbles, gravel, sands, and mud— into the American, Yuba, and Feather rivers, where some of the worst flood devastation took place in the Sierra. These sediments made their way downstream to the Sacramento River, filling the channel and thereby exacerbating the flooding in Sacramento. The sediment pulse eventually made its way into the Central Valley, delta, and San Francisco Bay.

Geologist G.K. Gilbert calculated that more than 1.5 billion cubic feet of mining debris filled the river systems before the practice of hydraulic mining was outlawed in 1882 in the first federal environmental case in history. Unfortunately, the mining debris continued to aggravate flooding in these rivers for the rest of the nineteenth and into the twentieth centuries, until it was finally flushed out of the system.

Out the Golden Gate

San Francisco Bay and its narrow outlet to the Pacific Ocean at the Golden Gate funnel all the rain and floodwaters from most of Central and Northern California. That wet winter, a steady flow of muddy, debris-laden water twenty feet deep could be observed pouring through the San Francisco Bay and out the Golden Gate for months. The sheer mass of all this freshwater acted as a wedge, pushing back Pacific seawater and preventing it from entering the bay, making the San Francisco Bay more like a freshwater lake. The enormous volumes of mud and silt carried by the turbulent waters into the bay filled shallow marshlands and caused a shoaling of the bay by up to three feet, causing the expansion of marshlands into the estuary and completely wiping out the thriving oyster industry that had developed around the shores of the bay since 1850.

We can only speculate about what was washed down to San Francisco Bay and out the Golden Gate during those epic floods along with the muddy waters: shredded homes, barns, trees, furniture and personal belongings, even the remains of people and animals. All were carried out to the Pacific and a watery grave.

The 1861–62 floods extended far beyond the borders of California: they were the worst ever in recorded history over much of western North America, including northern Mexico, Oregon, Washington, and British Columbia. Moreover, the rain-bearing storms did not stop in the Sierra Nevada but continued eastward into Nevada, Utah, and Arizona.

A normally arid state, Nevada received about twice its typical annual rainfall in the two-month period of December 1861 to January 1862. This excess water transformed the Carson Valley into a large lake. Villages and settlements, located on higher ground along the eastern slopes of the Sierra Nevada, were mostly spared. But at lower elevations, towns and cities suffered. Nevada City was inundated with nine feet of rain in sixty days.

In southern Utah, 1861–62 became known as the "year of the floods" as homes, barns, mills, and forts were washed away, including the adobe home of a Mormon bishop, John D. Lee. In his diary, edited by R. G. Cleland and J. Brooks, Lee had carefully recorded the weather throughout January 1862, noting a solid period of alternating rain and snow with strong winds for most of that month. In early February, as Lee attempted to move his ten wives and his children from their adobe fort that was disintegrating from the heavy rain, strong winds blew down part of the wall into a bedroom, killing two of the children.

Meanwhile, the village of Tonaquint in southwest Utah, situated at the confluence of the Santa Clara and Rio Virgin rivers, was destroyed in January 1862 as the floodwaters chased the villagers from their homes, leaving only mud and debris for miles. Missionaries had first arrived there in 1856 to help the Native Americans build an irrigation system and a dam. Brigham Young had visited Tonaquint in May 1861, just months before the destructive winter floods, and proclaimed that a city of spires, towers, and steeples would be built there. But before Young's vision could become a reality (the village later became the city of St. George), torrential rains deluged the region, engulfing southern Utah with rising waters. And just ten miles up the Santa Clara River to the northwest of Tonaquint, a group of Swiss Saints had settled in the village of Santa Clara, growing fruit, grapes, and cotton. When the floods struck, the Swiss Saints fled for their lives, their peaceful village utterly destroyed.

In Oregon, two and a half weeks of solid rain caused the worst flooding in that state's history. Deluges covered huge portions of the lower Willamette Valley where Oregon City is located. In 1861–62, Oregon City was the terminus of the Oregon Trail and the state's capital, where George Abernathy, an Oregon pioneer and the state's first elected governor, lived and ran a thriving business until the flood destroyed his home, forcing him (and many others) to leave.

Arizona, too, was affected: floods occurred in the Gila, Verde, Bright Angel, and Colorado river basins between January 19 and 23, 1862, and flooding was severe in Yuma, Arizona.

ADDING MISERY TO SUFFERING

The suffering endured by the thousands of homeless and stranded residents was not over after the rains. In fact, the situation worsened in late January, when unusual arctic conditions descended on the West Coast. The freezing weather struck from the Pacific Northwest to Southern California. In Oregon and Washington, the Columbia and other rivers froze. In California, Alcatraz Island in San Francisco Bay was ringed with ice, and snow covered the hills surrounding the bay. Six inches of snow fell in Napa, one inch of snow in Sacramento, and snow covered the entire northern Sacramento Valley. Temperatures in San Francisco dropped to 22°F, with frosts killing any remaining crops all the way to the Mexican border. In Red Bluff, 125 miles north of Sacramento, ice and snow covered the region. The freezing conditions were especially difficult for the thousands of homeless flood refugees across the region.

THE LESSONS LOST

Why so many people were caught off-guard by these floods remains puzzling. It appears that the Native American populations, who had lived in the region for thousands of years, had deeper insights into the weather and hydrology, and they recognized the patterns that result in devastating floods. A piece in the *Nevada City Democrat* on January 11, 1862, reported that Native Americans left Marysville for the Sierra foothills a week before the large flood, predicting higher water than at any time since the region had been settled by European-American pioneers.

The specific weather pattern that the Native Americans of the West recognized as bringing particularly severe flooding is finally understood again today—at least by climatologists. The powerful storms originate in the warm and moist tropical Pacific Ocean, near Hawaii; hence they have been nicknamed "Pineapple Expresses." Recent research describes these storms as "atmospheric rivers," and they often result in the worst floods in the West, particularly in California. When they follow on the heels of arctic storms that leave several feet of new snow on the mountains and freeze the soil (creating an impermeable surface), atmospheric rivers pack a strong meteorological punch that can be especially deadly. The warm rains, unable to penetrate the surface, rapidly flow over the surface, causing extreme runoff.

In 1861–62, however, immigrants did not recognize the climatic warning signs. They had never experienced such extreme flooding in the twelve years since the Gold Rush began, though lesser floods were not uncommon in the region.

That year, though many of the western states experienced their worst floods on record, California bore the brunt of the damage. The costs were devastating: one quarter of the state's economy was destroyed, forcing the state into bankruptcy. The tragic 1861–62 floods may have temporarily served to awaken the residents of California and the West to the possible perils of their region's climate. They saw nature at its most unpredictable and terrifying, turning in a day or an hour from benign to utterly destructive. But the costs to the state went beyond the loss of life, property, and resources; California's spirit and confidence were badly shaken.

The message to residents of the lowlands was clear: living in the large valleys of the West is risky, even deadly. These valleys were created over time by the rivers that continue to flow through them, and the broad, flat valley bottoms are called floodplains for good reason: the rivers periodically spill over their banks and spread across the broad expanses.

But this message has been lost in the hurley-burley growth during the century and a half since, as California's population, agricultural production, and technological innovation soar, transforming the state and its economy beyond all expectation. Development in California and the West proceeds with homes and towns built in exactly the same places that were obliterated 150 years ago. Residents now live in floodplains throughout the American West, despite the known risks to life and property. Perhaps modern westerners are not so different from their forebears, many of whom were pioneers or '49ers panning for gold. After the great 1861–62 floods, Brewer wrote of Californians: "No people can stand calamity as this people. They are used to it. Everyone is familiar with the history of fortunes quickly made and as quickly lost. It seems here more than elsewhere the natural order of things.

I might say, indeed, that the recklessness of the state blunts the keener feelings and takes the edge from this calamity" (pp. 249–50).

Nature in California is at its most magnificent and terrifying, with awesome tectonic forces deep in the earth producing both the steep and rugged beauty of the coast and the vast interior mountain ranges. Massive destruction from high-magnitude earthquakes comes with the territory, as does cataclysmic flooding.

A critical element of living in a place like California is an awareness of these natural disasters, and this requires a deep understanding of the natural patterns and frequencies of these events. We have building codes today for earthquake safety as well as city emergency plans. But are the millions of new westerners really aware of the region's calamitous climate history? Most have never even heard of the 1861–62 floods, and, as we will explore in the coming chapters, the 1861–62 floods may not have been the worst that nature can regularly dish out to the region. We will explore the evidence for similar, if not larger, floods that have occurred every one to two centuries over the past two millennia in California. But first, we take a look at nature's flip side with examples of the worst historic droughts of the twentieth century, including the Great Dust Bowl.

THREE

The Great Droughts of the Twentieth Century

> The rains disappeared—not just for a season but for years on end. With no sod to hold the earth in place, the soil calcified and started to blow. Dust clouds boiled up, ten thousand feet or more in the sky, and rolled like moving mountains—a force of their own.
>
> TIMOTHY EGAN, *The Worst Hard Time*

THE GREAT DUST BOWL DROUGHT was considered the worst climate tragedy of the twentieth century in the United States and the worst prolonged environmental disaster in its recorded history. The years between 1928 and 1939 were among the driest of the twentieth century in the American West and Midwest, with heart-breaking impacts stretching into the Midwest and Colorado, New Mexico, Oklahoma, Texas, and Kansas. At its peak in 1934, this drought affected three-quarters of the nation—stretching from the Great Plains north into the Canadian Prairie, and along the West Coast.

WHAT WENT WRONG?

On the Great Plains, unsustainable agricultural practices compounded the impacts of the drought on the land. For centuries, if not millennia, prairie grasses had stabilized the topsoil by retaining moisture beneath the surface through freezing winters and dry summers. Even in times of drought, the grasses provided food and shelter for snakes, jackrabbits, prairie chickens, and other wildlife.

During the 1920s, however, farmers on the Great Plains began removing the native grasses down to their roots. Initially, the effects were not apparent. But as millions of acres of prairie were converted to wheat production with dry farming on a colossal scale, the damage was done. The farmers knew

FIGURE 11. Dust storm approaching a midwestern town during the Dust Bowl in 1932. (Photo from the Natural Resources Conservation Service, U.S. Department of Agriculture.)

that the loose soils they were creating had nothing to stabilize them should drought return to the region, but they assumed the abundant rainfall of the early twentieth century would continue. Over a few short years, the prairie lands were ploughed so deeply and grazed so heavily that almost all the native grasses were gone and the land left vulnerable.

The West began its decline into drought in 1930. Many of the farmers held on, hoping the rains would return as they always did. But the drought continued unabated for eight years. In the early 1930s, strong winds began roaring over the prairies, creating a new and more visible catastrophe. The winds swept up topsoil from the Great Plains, forming thick clouds of dust, silt, and sand and darkening the skies—covering the earth like a blanket of dirty snow. The dust seemed to find its way everywhere: seeping under doors and through window cracks into the houses and into food stores, clothes drawers, beds, eyes, noses, and scalps.

The dust storms, or "black blizzards," blew across hundreds of miles, even reaching as far as Washington, D.C., New York, and New England. Some reports described the dust falling on ships sailing across the Atlantic Ocean, and, when lofted high enough to be entrained by the jet stream, the dust reached as far away as Europe.

At the peak of the Dust Bowl, dust settled over nearly three-quarters of the United States. An estimated 100 million acres were destroyed or debilitated to varying degrees. The residents of the Great Plains could tell where the dust originated by its color: red dust came from eastern Oklahoma, yellow-orange dust from Texas, and black dust from Kansas (see figure 11).

BLACK SUNDAY

April 14, 1935, is remembered as Black Sunday, when a particularly bad dust storm blew in from the north. In Kansas, the day started out sunny, clear, and windless—a brief hiatus from the dusty clouds that had blown through the region regularly the previous four years. Lulled by this seemingly quiescent day, the Kansas residents began removing the sheets that covered their windows—opening them up to let in the rare fresh air. They walked outside without their usual protective aids: goggles, homemade sponge-masks over their mouths, petroleum jelly in their nostrils. This spring day seemed to be a turning point—a new start. As the day wore on, however, people spotted enormous black clouds moving in from the north, and some momentarily thought a rainstorm was headed their way. But they soon learned that these were not ordinary storm clouds, dark with water: these clouds were blackened by silt and sand.

A towering wall of earth, 2,000 feet high and 200 miles wide, soon plunged them into darkness. So thick and dense was the dust that people could not see their own hands in front of them, or their house just a few feet away. Many were blinded and suffocated as the wind blew dust into their eyes and lungs. The cold, turbulent winds moved south at 80 miles per hour, entraining soil and dirt and throwing it aloft as it traveled across South Dakota and into Nebraska, Kansas, Colorado, Oklahoma, and the Texas panhandle.

After years of breathing dust, residents of the Great Plains suffered respiratory ailments such as sinusitis, laryngitis, bronchitis, and even pneumonia and tuberculosis. Children and the elderly suffered severe chronic coughing and chest pain, shortness of breath, and nausea. Although the Red Cross opened emergency hospitals to help deal with this medical crisis, impassable roads covered with dust drifts made it impossible for many people to reach them.

As the Great Plains became increasingly uninhabitable throughout the 1930s, many residents were forced to leave. Banks foreclosed the homes and farms of millions of people, driving hundreds of thousands of the homeless and desperate farther west to places like Arizona, Oregon, and California.

California and other neighboring states were facing hard times of their own. This region was suffering the longest and most severe drought on record, as measured by rapid drops in lake levels, lower precipitation, and reduced river flows across the West. Samuel Harding, a University of California scientist, compiled these data in a 1965 report, and he noted that many western states—including California, Arizona, Nevada, and Oregon—had enjoyed two very wet decades in the early twentieth century. But the climate took a dry turn after 1925, when rainfall and runoff began to steadily decline until the late 1930s.

In California, the flows of major rivers like the Sacramento, Feather, and Kern dropped below average levels from 1927 to 1935. At the same time, Lake Tahoe—the largest alpine lake in North America—fell fourteen feet below its sill depth to a historic low. The lake's only outlet, the Truckee River, experienced its lowest flow rates on record during this period, finally ceasing to flow altogether. Downstream, low river flows led to significant saltwater intrusion into the Sacramento–San Joaquin Delta between 1931 and 1939, causing record high salinity levels in both the delta and San Francisco Bay as well as declines in aquatic ecosystems.

To the north, at the border of California and Oregon, the level of Goose Lake steadily dropped between 1926 and 1931 before the lake finally desiccated in 1936. Similar declines were seen in closed-basin lakes from eastern Oregon to central British Columbia. Farther east, the Great Salt Lake in Utah also dropped to its lowest level on record during 1934 and 1935, and, in the Southwest, Colorado River flow measured at Lees Ferry, Arizona, fell below its long-term average between 1930 and 1940. Looking at all the evidence, Harding concluded in his report that the period from 1927 to 1935 was the driest of the previous 200 years.

The Dust Bowl drought years were cause for grave concern to water planners and residents in Southern California as well, where lower than average river flows were measured on the Santa Ana and San Gabriel rivers. The newly constructed Los Angeles Aqueduct from the Owens Valley to Southern California allowed the population of Los Angeles to grow from 500,000 to

1.2 million between 1920 and 1930. The lingering drought caused water planners to doubt whether the new aqueduct would be sufficient to supply the growing city into the future. In a vicious cycle, the aqueduct allowed the region's population to grow, but the larger population required still more water.

As the Dust Bowl drought continued unabated, Los Angeles voters felt they had no choice but to approve a bond measure to extend the Los Angeles Aqueduct farther north, into the Mono Basin. In the 1930s, the city of Los Angeles built diversion dams on four of the five tributary rivers that fed Mono Lake. As this water was diverted south to the growing metropolis over the next four decades, Mono Lake shrank steadily.

TAPPING ANCIENT GROUNDWATER

Meanwhile, a third of a million Dust Bowl refugees made their way to California, settling in the southern two-thirds of California's Central Valley—a region known as the San Joaquin Valley. At the time, the drought was threatening to curtail California's developing agricultural industry. However, powerful electric pumps, which became available to farmers for a reasonable price in the 1930s, allowed farmers to tap an extensive source of water that had been accumulating beneath the ground for thousands of years. These ancient aquifers had already been used to irrigate approximately one and a half million acres in the San Joaquin Valley by 1930, transforming that region into the "bread basket of the world"—or, more precisely, the "fruit and nut basket of the world."

During the Dust Bowl drought, Central Valley farmers relied on this vast reservoir of groundwater for their crops—water that used to bubble to the surface as natural springs. Groundwater was pumped so extensively that the water table dropped dramatically, in some places by up to three hundred feet, killing deep-rooted oak trees that were several hundred years old. At the rate it was being pumped, this vast underground water reservoir would be depleted in only thirty to forty years.

Agriculture was becoming California's top industry, and, in the face of the Dust Bowl drought and rapidly depleting groundwater, an immense water project was planned. The goal of this project was to capture rain and snowmelt draining from the Sierra Nevada mountain range behind dams and then to transport the water in reservoirs via pipes and aqueducts from the moist northern half of the state to the much drier southern half, including the burgeoning San Joaquin Valley agricultural lands and growing cities in Southern California. A massive water conveyance canal, called the California Aqueduct, was completed during a relatively wet period in the state that followed the Dust Bowl drought, in the 1940s and 1950s.

THE DROUGHT OF 1987–1992

The Dust Bowl drought was one of only two droughts that have lasted more than four consecutive years in California since 1850, when the state began keeping records. A later drought of similar magnitude and duration began in 1987, and it makes for an interesting comparison. California's population had expanded from seven million in the 1930s to thirty-one million in the 1980s, and agriculture had grown to an industry worth $30 billion a year, heavily dependent on the extensive water infrastructure that had been built during the intervening years.

During the multiyear drought that began in 1987, annual precipitation over much of California was down to only 50 percent of the twentieth-century average. This drought affected the entire state, and, though no single year was as severe as the 1976–77 drought described in chapter 1, the cumulative effects were ultimately more devastating.

The 1987–92 drought crept up on the state. Reservoirs had enough water stored to buffer the first three to four years, but, after the fourth year, reservoir storage statewide was down 60 percent. Severe restrictions, with reductions of 75 percent, were imposed on water delivery for agricultural uses, and reduced river flows caused a decline in hydroelectric power generation.

In Southern California, the two reservoirs for the city of Santa Barbara dried up almost entirely, forcing the city to build a pipe connecting to the California Aqueduct through the Santa Ynez Mountains at a cost of hundreds of millions of dollars. Many more millions were spent by the city on one of the world's largest seawater desalination plants, built as an insurance policy for the vulnerable city.

The impacts on the environment were also mounting. Bark beetle infestations caused widespread mortality in the forests of the Sierra Nevada, where extensive stands of pine trees, already weakened from the drought, succumbed to disease. Reduced stream flow led to major declines in fish populations, including the chinook salmon and striped bass, decimating commercial and recreational fisheries.

People sought new sources of water as, year after year, the drought continued. The use of groundwater became especially popular: private well owners deepened existing wells or drilled new ones, and thirsty consumers in the San Joaquin Valley extracted eleven million acre-feet more groundwater than could be replenished naturally, further lowering the water table. Cloud-seeding operations were expanded in the Sierra Nevada in the hopes of increasing precipitation, even if only by a fraction of a percent.

Overall, the 1987–92 drought resulted in financial losses of $3 billion to the state of California. This was an even greater financial disaster than the drought of the 1930s had been. The state's population had expanded by over four times what it had been earlier in the century, and a burgeoning agricultural industry relied more heavily on water from reservoirs.

California has not experienced as severe a multiyear drought since 1992, but the population of the state continues to grow and is predicted to double by the year 2050. Moreover, California faces an uncertain climatic future: global warming is predicted to change the state's hydrology, with less snowpack and overall warming and drying, more frequent flooding as more winter precipitation falls as rain rather than snow, and an increase in major storms. California, like the rest of the American West, also faces the recurrence of more severe climate events. This will be discussed in the coming chapters.

CLIMATE: THE LONG VIEW

The extreme climate events of the past century and a half that have been presented in this chapter were indeed severe. However, paleoclimate evidence suggests that they were by no means the worst we can expect from nature. In fact, compared to what records indicate of the long-term climate patterns for this region, the climate of the North American West over the past 150 years has been nearly ideal for human settlement.

We are now learning that the floods and droughts of this period are an incomplete sampling of the extreme events that have been a "normal" part of the West's hydrology for thousands of years. The clues from past climates depict extremely dry periods—in some cases lasting decades or even centuries—often punctuated by torrential rains and floods. We now realize that vastly larger floods and more severe and protracted droughts, though rare, are as inevitable as the earthquakes and volcanic eruptions that Pacific Coast residents know will occur.

But most westerners are unaware of these prehistoric extreme climate events that complete the story of the region's long-term climate pattern. The evidence has been growing, but only recently have the clues been pieced together to form a more accurate picture. This book presents an outline of what we know about the past 20,000 years, from the peak of the last ice age to the present. During those millennia, climate has often varied by extremes in the American West. Close examination of the evidence suggests that the benign conditions of the past century and a half have not prepared us adequately for what could come in the region's climatic future.

The droughts of the 1930s and 1987–92 are believed to have originated in the Pacific Ocean. During these periods, sea surface temperatures in the tropical Pacific were unusually cold, which is characteristic of the ocean-atmosphere phenomenon known as La Niña. The next chapter explores the connections between the Pacific Ocean and the American West and how extremes of wet and dry are products of the interactions between this enormous body of water and the atmosphere above it.

FOUR

Why Is Climate So Variable in the West?

> El Niño has taught two lessons that will endure. The first is that
> large-scale variability such as El Niño is not a disaster, anomaly,
> or cruel twist of fate; it is how Earth works. To mature and live
> harmoniously in the Earth system, human culture must adapt
> to Earth's rhythms and use natural variability to its advantage.
>
> MICHAEL GLANTZ, *Currents of Change*

CLIMATOLOGISTS IN THE AMERICAN WEST have been carefully recording daily observations of weather conditions for the past 150 years. They have also monitored ocean conditions over recent decades, providing insights into the critical interactions between the Pacific Ocean and its overlying atmosphere. In this way, climate scientists have assembled many seemingly unrelated observations into a larger picture that explains, at least in part, why climate in the American West changes from year to year, decade to decade, and perhaps over even longer timescales. This chapter provides an overview of the discoveries that begin to explain why climate is so variable in the West—at least over the past century and a half.

WHERE WESTERN CLIMATE PATTERNS START

To understand what controls climate in the American West, we look first to the region's neighboring body of water, the Pacific Ocean, which is immense by any measure: it contains half the water on Earth and covers over one-quarter of the globe. In physical terms, the ocean is a vast heat engine as well as a source of water for the western United States. Although the ocean is 2.5 miles deep (on average), the temperature of the waters in just the top 300 feet or so is what affects climate in the West by altering large-scale patterns of atmospheric pressure and circulation, which are factors in the size and movement of rain-bearing storms.

The two pressure cells that most influence climate over the American West are the "Pacific High," which extends from just east of Hawaii to the West Coast, and the "Aleutian Low," which extends from the Gulf of Alaska to the Aleutian Islands in the North Pacific. The basic physics of our spinning planet result in the circulation of upper-level winds in a counterclockwise direction around the Aleutian Low and in a clockwise direction around the Pacific High. The Aleutian Low is strongest during the winter, when the Pacific High is weakened and displaced to the south. In the mid-latitudes— roughly midway between the equator and the North Pole—movement from west to east of air masses that form over the Pacific creates the prevailing westerly wind pattern over the West. In wintertime, most of the rain-bearing air masses bring moisture from the ocean across the northern coastal states (Washington, Oregon, and northern California).

Much of the moisture falls as the air masses are forced up the slopes of first the coastal mountains and then the Cascade or Sierra Nevada mountain ranges. In the higher elevations, this precipitation often falls as snow. In the northern Pacific Ocean, the Aleutian Low occasionally moves farther west, and thus a high-pressure cell may form over the eastern Pacific just off the West Coast, preventing storms from entering the West. This situation is often associated with drought.

During the summer, the Pacific High strengthens and moves farther north off the West Coast, deflecting storms into the Gulf of Alaska and Canada. California is positioned in just such a way that it lies under the eastern side of the Pacific High during the summer months, which effectively blocks storms from moving into the state. For this reason, California is one of the few regions in the world that receives virtually no summer rain.

In contrast, the southwestern region of North America—including Arizona, New Mexico, and Colorado—receives moisture in the summer during what is called the "North American Monsoon." The monsoonal rains occur when moist air from the Gulf of Mexico, the Gulf of California, and the eastern Pacific Ocean moves northward into the southwestern United States, which has undergone intense solar heating during the months of July, August, and September. Monsoonal rainfall can account for as much as half of the annual precipitation in the U.S. Southwest, although it is highly variable since the region lies at the northern border of the monsoon (with the core located over northern Mexico).

Precipitation in the West is highly variable from year to year and decade to decade. Research over the past few decades has revealed that these seemingly random and unpredictable swings in climate are largely brought about by periodic changes in the Pacific Ocean and the atmosphere above it, producing phenomena known as the El Niño–Southern Oscillation (ENSO) and the Pacific Decadal Oscillation (PDO). Together, the ENSO and the PDO are responsible for climate patterns over most of the western United States, including precipitation, wind patterns, and air temperature during various seasons. The ENSO arises from conditions in the tropical Pacific Ocean; the PDO is tied to conditions over portions of the northern Pacific Ocean. ENSO events are typically six to eighteen months in duration, whereas PDO phases last two to three decades or more.

THE EL NIÑO–SOUTHERN OSCILLATION (ENSO)

Conditions in the tropical Pacific Ocean and in the air above it play a major role in the year-to-year variability of climate in the North American West, and they are often responsible for the region's climate extremes.

The earliest scientific observations leading to our present understanding of oceanic and atmospheric variability in the tropical Pacific were made in the early twentieth century by Sir Gilbert Walker, director of the Indian Meteorological Department. Walker's expertise was in mathematical physics and statistics, not meteorology, but he had a passion for analyzing a wide array of seemingly unrelated meteorological data to explain such phenomena as droughts in India. He observed that there was a "swaying of pressure on a big scale backward and forward between the Pacific Ocean and the Indian Ocean" (p. 22). Walker called this seesawing surface pressure the Southern Oscillation, which is part of an east-west atmospheric cycle now known as the "Walker Circulation."

Later in the twentieth century, Norwegian meteorologist Jacob Bjerknes, who later founded the Department of Meteorology at the University of California, Los Angeles, observed oceanic and atmospheric patterns in the tropical Pacific Ocean during the strong 1957–58 El Niño (and during subsequent events in 1963 and 1965). These conditions, characterized by anomalously warm ocean surface waters in the eastern tropical Pacific that occur in late December, have been recognized for centuries by Peruvian fishermen, who noticed significant reductions in their catch. The episodes studied by Bjerknes were also characterized by light easterly trade winds, heavy rainfall in the eastern Pacific, and the atmospheric pressure shifts described by Walker.

In the late 1960s, Bjerknes brought together all of these disparate pieces of information into a hypothesis that seemed to explain the entire ocean-atmosphere phenomenon. He proposed that, in typical years, the constant east-to-west (or easterly) trade winds push warm surface waters across the Pacific Ocean to the western end, where they pile up in a region called the "Pacific warm pool." This "pool" of warm water is roughly four times the size of the continental United States, measuring 9,000 miles from east to west and 1,500 miles from north to south. High rates of evaporation off this body of warm water provide abundant rainfall for places such as Australia and Indonesia.

On the opposite side of the Pacific, just off the coasts of North and South America, the surface waters blown to the west by the trade winds are replaced by colder water from beneath the surface in a process known as "upwelling." These cold, upwelled waters cool the overlying atmosphere, forming high pressure, resulting in very dry conditions in the adjacent land areas, including Ecuador, Peru, and Chile to the south and the American Southwest to the north.

El Niño Episodes

This vast, complex ocean-atmosphere system periodically stumbles, however, when trade winds mysteriously falter or even fail completely for months, producing the pattern known as El Niño. This phenomenon, which means "the child," was named in South America, where it usually appears around Christmas and thus the celebration of the birth of the Christ child. During El Niño, the western Pacific warm pool—elevated several inches higher than the eastern Pacific—is released from the invisible force of the trade winds that hold it, and it flows eastward as a low-amplitude—but tremendously long—wave (called a "Kelvin wave") across 8,000 miles of the Pacific Ocean toward the Americas at about 150 miles per day.

Upon reaching the eastern Pacific, this wave splits and travels along the coasts of North and South America, causing sea levels to rise by about six to ten inches—the amount equal to the height of the wave. Higher sea levels can lead to coastal flooding and erosion.

The wave of warm water also acts as a cap, suppressing the colder, nutrient-rich waters below. Cold water is denser and thus unable to rise above the warmer waters. These atypically warm surface waters pump water vapor, heat, and enormous amounts of energy into the atmosphere above, creating a column of warm, moist air. This column bulges into the atmosphere for six to ten miles, where it remains like a giant boulder in a stream—an obstacle that atmospheric winds and storm systems must move around—and so influences the strength and location of jet streams over the northern and southern Pacific Oceans.

The copious amounts of evaporation from this warm water also spawn large storms that are carried by westerly winds into northern Mexico and the southwestern United States. The deluges, floods, and mudslides experienced in the North American Southwest are often mirrored by opposite but equally tragic events in distant regions thousands of miles away, on the western side of the Pacific Ocean. There, the surrounding ocean waters are cooler than usual, with less evaporation and less rain. Droughts, forest fires, and dust storms often occur simultaneously in places such as Australia and Indonesia. The Pacific Northwest also experiences drier than normal conditions during an El Niño event.

La Niña Episodes

If El Niño conditions represent a disruption of what we think of as the normal climate state, La Niña episodes occur when "normal" conditions go into "overdrive." The trade winds blow even harder across the Pacific Ocean, forming an even larger warm pool in the western Pacific and leading to an increased upwelling of cold surface waters in the eastern Pacific. In the western Pacific, evaporated water off this vast warm pool rises high into the atmosphere, forming towering clouds that drench the tropical regions over northern Australia and Indonesia. Across the Pacific, in Peru and in the North American Southwest, the unusually cool waters off the eastern Pacific bring severe drought.

ENSO and Precipitation in the West

Because the oceans provide the water and energy for most of the storms on Earth, the shifting of warm surface waters to the central and eastern Pacific during El Niño ultimately alters precipitation patterns across much of the globe by atmospheric teleconnections (or long-distance relationships between weather patterns). For example, droughts in Australia, Indonesia, southern Africa, southern India, Spain, and Portugal often occur at the same time as deluges and floods in Peru, the Mississippi River Basin, and Western Europe. In the American West, the Southwest normally experiences deluges

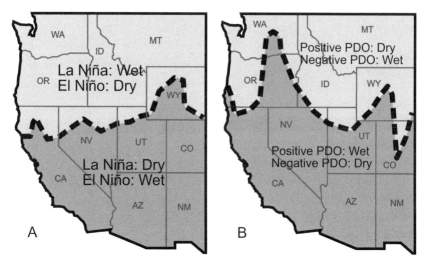

FIGURE 12. Maps of the western United States showing (A) the correlation between winter (November through April) precipitation and the El Niño–Southern Oscillation, and (B) the correlation between winter precipitation and the Pacific Decadal Oscillation. (Maps courtesy of Dr. James Johnstone at the Joint Institute for the Study of the Atmosphere and Ocean, University of Washington.)

and floods during El Niño while the climate in the Pacific Northwest is unusually dry (see figure 12A). The geographic transition between these two opposite climate responses often occurs in Central California, which can experience extreme conditions in either direction.

Together, El Niño, La Niña, and the Southern Oscillation are major players in the ENSO, that ocean/climate phenomenon of interrelated (and often destructive) shifting pressures, wind-ocean currents, and ocean surface temperatures. The Southern Oscillation Index (SOI) is a measure of the differences in atmospheric pressure between the western and eastern tropical Pacific Ocean (measured in Tahiti and in Darwin, Australia), and it reflects the state and strength of the system. When the SOI is negative, air pressure is below normal in Tahiti and above normal in Darwin (with warmer waters in the eastern Pacific)—typical of an El Niño year. During periods of negative SOI, precipitation is highest in the Southwest and lowest in the Pacific Northwest. A positive SOI brings the opposite climate patterns: dry in the American Southwest, wet in the Pacific Northwest.

The link between the ENSO and precipitation in the western United States arises, in part, from the location of the jet stream, which is a corridor of fast-flowing air occurring at the boundary between cold air to the north and warmer air to the south. Low pressure systems tend to form beneath the jet stream and move along with it, bringing wet storms from the Pacific Ocean into the western United States in the winter. During El Niño events, the warm surface waters in the eastern Pacific lead to a greater temperature contrast farther south, pulling the jet stream southward and bringing more storms into the Southwest. In contrast, during La Niña events, the eastern Pacific Ocean becomes even cooler than normal, reducing the temperature contrast and therefore reducing the likelihood of a jet stream into this region, decreasing the occurrence of storms and precipitation in the Southwest.

The 1997–1998 El Niño: Biggest on Record

El Niño events recur every two to seven years, at varying degrees of severity. The 1997–98 El Niño was the most heavily monitored and closely observed climate event in history, and it taught climate scientists much of what they know about the ENSO today. At the start of this event, unusually warm waters began moving eastward across the equatorial Pacific during the spring of 1997, alerting major oceanographic and atmospheric research institutions that a massive El Niño event was on its way. Researchers worldwide watched as the wave of warm ocean waters made a rapid journey eastward across the Pacific, warming oceanic islands and coral reefs all along the way and reaching the west coast of South America by August of that year.

Oceanographers and climatologists had already deployed an array of buoys across the tropical Pacific Ocean after the monstrous 1982–83 El Niño event had taken them by surprise. These buoys monitor water temperature between the surface and a depth of 1,600 feet, wind speed and direction, air temperature and humidity, and water movement (currents). The data are automatically relayed back to computers via the Argos satellite system and are used to predict an upcoming El Niño event several months in advance, which is the amount of time it takes for tropical ocean currents to respond to changes in the trade winds.

In 1997, scientists were predicting deluges in Ecuador and Peru, California, the Southwest, and throughout the southern United States, and they expected droughts in the American Midwest, Brazil, Australia, and Indonesia. Their predictions were spot on, although the actual impacts were significantly greater than anticipated. Storms in November 1997 generated massive flooding in southern Ecuador that left over three thousand people homeless in minutes. Floodwaters also swept through coastal cities and agricultural lands in Peru. By December, an area of warm water greater than the size of the United States covered the eastern Pacific, heating the overlying air from the eastern Pacific to Central America and even into the Atlantic from Cuba to Libya, imposing its own weather patterns. The Northern Hemisphere jet stream shifted in February 1998, bringing warm tropical air northward and warmer than average temperatures into North America, Europe, and eastern Asia.

Half a world away, rice paddies in the Philippines were dry and dying, with more than $20 billion in lost crops, food shortages, and several hundred deaths. Drought led to catastrophic wildfires in Indonesia because the normal monsoon rains never came, and the fires filled the air with thick clouds of smoke, causing rampant respiratory illnesses. These disasters occurred despite an early warning system and programs that gave advance notice of droughts, floods, and potential crop failures to affected regions of the world. The 1982–83 and 1997–98 El Niño events combined caused $120 billion in damages and 24,000 deaths worldwide.

THE PACIFIC DECADAL OSCILLATION (PDO)

Scientists have been monitoring the ocean and atmosphere over several decades and have detected long-term climate patterns originating in the North Pacific Ocean. One of these patterns was discovered when fisheries scientists Robert Francis and Steven Hare began analyzing cycles in salmon populations off the coast of Alaska in the early 1990s. When they compared the salmon population sizes with climatic and oceanographic data over a period of several decades, they realized that the two cycles appeared to be tracking each other.

Climatologist Nate Mantua and colleagues at the University of Washington further analyzed the data and correlated fluctuations in salmon productivity in the area between Alaska and the Pacific Northwest with changing ocean conditions during the twentieth century. The team observed that changes in wind direction during wintertime influence the temperature of surface waters along the Pacific Coast of North America. When the direction of the winds causes increased upwelling of cool, nutrient-rich waters, the enhanced growth of phytoplankton and krill allows salmon populations to increase. During periods of decreased upwelling, sea-surface temperatures rise and nutrient levels are lower, causing krill and phytoplankton to decrease, leaving less food for salmon and thus reducing their populations.

Mantua and his colleagues analyzed the oceanographic and fisheries data and found that these cycles of sea-surface temperature and salmon populations lasted decades. They discovered that sea-surface temperatures in the northeast Pacific were warmer than average from 1925 to 1947 and from 1977 to 1999, and that they were cooler than average from 1947 to 1977 and after about 1999. The team called these broad oceanographic cycles the Pacific Decadal Oscillation, or PDO. The warm and cold phases each lasted two to three decades.

The PDO is an elegant demonstration of the linkages between climate, oceans, and marine life, and it also helps explain the enigmatic shifts that occur every few decades in the West's climate as measured by rainfall and streamflow across the region. During a positive (warm) phase of the PDO, surface waters of the central North Pacific are anomalously cold while surface waters along the northwestern coast of North America are unusually warm. There is also warming of tropical surface waters in the eastern Pacific. This PDO phase brings higher than average winter precipitation to the American Southwest, with increased risk of flooding, while lower than average winter precipitation occurs in the Pacific Northwest. The patterns reverse during a negative (cool) PDO phase, with cooling of surface waters along the northwestern coast of North America and in the tropical eastern Pacific. During this phase, average winter precipitation is lower in the American Southwest and higher in the Pacific Northwest (see figure 12B).

Two full PDO cycles have occurred in the past century: cool, or negative, regimes occurred from 1890 to 1924 and from 1947 to 1976; warm, or positive, regimes occurred from 1925 to 1946 and from 1977 to 1998. Regime shifts— transitions between positive and negative phases of the PDO—occur every few decades and can also have dramatic effects on the climate. For instance, the winter of 1976–77 marked the transition from a cool to a warm phase of the PDO, and, during that year, extreme drought had societal impacts beyond temporary water restrictions. It was the driest year on record in California, the Pacific Northwest, and other regions of the North American West. In the Pacific Northwest, the human population and agriculture had expanded during the wetter, cool-phase PDO between 1945 and 1976. The 1977 drought introduced the first of several dry decades in the region—and the first of many conflicts between farmers, fishermen, and other water users accustomed to wetter times.

Some of the deepest droughts and most catastrophic floods are caused by the interactions between the PDO and the ENSO. When the PDO is in a positive (warm) phase, El Niño events bring more rain to the American Southwest and less to the Pacific Northwest, as was the case during the period between 1977 and 1999. In contrast, during a negative (cool) PDO phase, La Niña events bring drier conditions to the American Southwest and wetter conditions to the Pacific Northwest, such as during the period from 1945 to 1976.

A negative (cool) PDO regime began in 1999, marked by a deep drought in the Southwest. But this time, the cold-phase PDO lasted only four years, flipping back to a warm phase from 2002 to 2005. It then returned to a cool phase again, where it has remained until the present time (2013). These recent fluctuations in the PDO are not typical for the twentieth century and are currently under investigation by atmosphere and ocean scientists to assess whether they are possibly related to global warming.

ATMOSPHERIC RIVER STORMS

Water Resources

Scientists at the U.S. Geological Survey and elsewhere have found that water vapor from the tropical Pacific Ocean is carried to higher latitudes through narrow corridors of concentrated moisture, dubbed "atmospheric rivers." Satellite technology over the past decade has allowed researchers to observe and image these storms. They discovered currents in the lower atmosphere (about one mile high, 250 miles wide, and extending thousands of miles) carrying water vapor and warm air from the eastern North Pacific to the West Coast. Once these atmospheric rivers hit the mountain slopes, they are forced upward and cooled, releasing copious amounts of precipitation along the Coast Ranges, Sierra Nevada, and Cascades.

Atmospheric rivers are actually a type of cyclone that forms at higher latitudes than hurricanes. Although they lack a cyclone's characteristic circular pattern, they carry hurricane-strength winds and a similar amount of rainfall. Unlike hurricanes, which obtain their energy from surface ocean heat content, these atmospheric rivers derive their energy from the contrasts in temperature between the tropics and the poles. Atmospheric river storms contribute between 30 and 50 percent of the total amount of precipitation that falls each year in California, so they are now recognized as a crucial source of water for the state. They also contribute a significant amount (10 to 40 percent) of the total annual precipitation in other western states, including Oregon, Washington, Nevada, and Idaho.

Floods

Atmospheric river storms are a double-edged sword: they provide critical water resources to the region, but they also produce the largest and most destructive floods along the West Coast. Climatologist Mike Dettinger of the U.S. Geological Survey and his colleagues have studied the past sixty years of climate and flood history in California and have shown that the years with the largest "Pineapple Express" storms (a type of atmospheric river storm that draws heat and vapor from the tropics near Hawaii and transports it to the West Coast) correlate with the largest floods. Dettinger also observed that these storms are tied to both the tropical Pacific (ENSO events) and the North Pacific (PDO events). His analysis reveals that, during El Niño when the PDO is in a positive (warm) phase, a higher proportion of precipitation in south-central California and western Washington comes from Pineapple Express storms.

Atmospheric river storms bring high volumes of precipitation—the equivalent of ten to fifteen Mississippi Rivers—from the tropics to the mid-latitudes over relatively short periods of time, and, because they are warm, more of that precipitation falls as rain high in the mountains rather than as snow, leading to severe floods. On the West Coast, atmospheric river storms occurred during the stormiest years on record—1861–62, 1997, and 2006—and other stormy years in between (1937, 1955, 1964, 1969, 1986, and 1994). The relationship between unusually warm storms and severe flooding was observed as early as 1900 by John Muir, who wrote: "The Sierra Rivers are flooded every spring by the melting of the snow as regularly as the famous old Nile. Strange to say, the greatest floods occur in winter, when one would suppose all the wild waters would be muffled and chained in frost and snow. . . . But at rare intervals, warm rains and warm winds invade the mountains, and push back the snow line from 2000 ft to 8000 ft, or even higher, and then come the big floods" (chap. 11).

Atmospheric river storms are likely responsible for the 1861–62 floods, according to scientists at the U.S. Geological Survey. But just how common are these storms and resulting floods? Was this a freak event that is unlikely to occur again, or is it something for which we should prepare? Millions of current westerners have never even heard of those floods, much less have any idea about the potential for another, similar event.

Only by examining the climatic history of the West preserved in its sediments, trees, landforms, and lakes can we begin to understand whether these large flood events have been repeated over centuries or millennia, and, on the opposite end of the climate spectrum, how large and frequent droughts were in the past. In the next few chapters, we will explore the evidence revealing past climate change in the West, including the startling evidence that the 1861–62 floods and the twentieth-century droughts may not have been the largest that nature can deliver.

PART TWO

A Climate History of the American West

Humanity has been at the mercy of climate change for its entire existence. Infinitely ingenious, we have lived through at least eight, perhaps nine, glacial episodes in the past 730,000 years. Our ancestors adapted to the universal but irregular global warming since the end of the Ice Age with dazzling opportunism. They developed strategies for surviving harsh drought cycles, decades of heavy rainfall or unaccustomed cold; adopted agriculture and stock-raising, which revolutionized human life; founded the world's first preindustrial civilizations in Egypt, Mesopotamia, and the Americas. The price of sudden climate change, in famine, disease, and suffering, was often high.

BRIAN FAGAN, *The Long Summer*

FIVE

Reading the Past

THE EARTH'S HISTORY BOOKS

> If you want to learn about nature, to appreciate nature, it is necessary to understand the language that she speaks in.
>
> RICHARD FEYNMAN, *The Character of Physical Law*

ARCHAEOLOGISTS DIG UP CLUES LEFT by ancient societies—bits of broken pottery, bones, beads, chipped stones—by sifting through ancient garbage dumps and other remains of their daily lives. By methodically sifting through and analyzing the preserved remains, they can infer much about past civilizations, their daily lives, and rises and falls. Similarly, paleoclimatologists learn about the earth's climatic history, including patterns of wet and dry, warm and cold, by looking for clues buried in glaciers, trees, fossils, corals, and sediments preserved in lakes, marshes, and the ocean.

To begin exploring what the natural climate archives have to tell us about the past, it is necessary to introduce a few of the ways that earth scientists decipher the clues of climate conditions in bygone eras. The story of the earth's ancient climate is not written in paper and ink: it is written in sediment, stone, trees, ice, and other types of natural materials, waiting to be uncovered, deciphered, and interpreted. Although the language is not familiar to most humans, earth scientists have devised ways, using increasingly advanced technology, to read and understand at least the basic grammar.

Paleoclimatologists use whatever evidence they can find to piece together past climate conditions, much like investigators at a crime scene. They search for clues to the climate that may have existed in the past and to ways past climate fluctuations may have affected the environment, ecosystems, and past human societies. Paleoclimate researchers have no access to direct records; no one was measuring and recording the daily temperatures and rainfall during prehistoric times. Instead, they use proxy—or indirect—evidence to assess conditions of the distant past. Examples of proxy evidence are mud from the bottom of lakes and ponds, microscopic organisms living in the oceans, bubbles frozen in glaciers, pencil-thin wood cores drilled from trees, and salts precipitating in dried-up lake bottoms. Much can also be learned about early climates through such straightforward clues as landforms: the heights and shapes of mountains, the locations and sizes of lakes and rivers, the locations of trees and other types of vegetation, and the sediments that accumulate at the bottom of lakes and the mouths of rivers.

We begin our exploration of the earth's natural history with one of the most ancient living natural archives: the bristlecone pine. These hardy trees, found throughout the western United States, are the oldest living creatures on the planet, and each season of their lives is recorded in their trunks. They retain a longer "memory" of the environment in which they grew than any other organism on Earth.

TESTIMONY OF THE BRISTLECONE PINE

On the harsh slopes of the White Mountains above Owens Valley in eastern California, a trail leads up to a forest of stunted pine trees. It is 1957, and a group of backpackers, loaded with an assortment of strange equipment, hikes up the trail. Douglas Powell (who told this story to his close friend, Frances Malamud-Roam, one of this book's authors) is the youngest of the hikers, and he finds it challenging—with the thin air and steep climb making breathing difficult—but, for the older members of the group, the hike is grueling. Powell arrives first at the lower reaches of a grove of scraggly trees, out of breath but excited. He knows that here in this ragged forest, perched at almost 11,000 feet, grows the oldest living tree on Earth. He surveys the trees around him while he waits for his companions, wondering which of these gnarled trees is "The One."

Standing at the forest's edge, Powell reflects on the harsh environment. Almost 8,000 feet above the Owens Valley floor, the atmosphere is thin, water is scarce, and the soil is rocky. The forest is sparse, with only patches of stunted, half-dead trees. The younger, smaller trees look like ordinary pines, though not as tall. The tough living conditions, harsh climate, and dry soils prevent the trees from growing very large. The older trees are more contorted, and only parts of the trunks appear to be living. Some trees appear to be all but dead (see figure 13).

FIGURE 13. A bristlecone pine growing in the White Mountains, eastern California. (Photo by B. Lynn Ingram.)

On the way up the trail, Powell had listened to his fellow hikers as they talked about the odd but hardy bristlecone pines. This group of scientists is on a research mission led by Edmund Schulman, who first discovered the surprisingly great age of these remarkable trees. For many years after this summer field trip, Powell remembered how Schulman, not yet fifty years old, sounded at times almost obsessed with the longevity of the trees. There on the slopes he observed them: the most gnarled were still alive at over 4,000 years of age.

Bristlecone pines are found throughout the Great Basin, from Colorado to California, but the oldest live only in the White Mountains of California, a range that runs parallel to and east of the Sierra Nevada range, just north of Death Valley. The White Mountains are high—up to 14,000 feet above sea level—and cold, windy, and dry. The range receives less than twelve inches of precipitation each year, most of it as winter snow. But even with such adverse conditions, the bristlecone pine survives and reaches extremely old age. Subsequent research has shown that the harsh conditions and extreme age of the trees are related. The slow growth rate of the bristlecone pine—with 100 tree rings in one inch of trunk—means that its bark is extremely dense, making it resistant to rot, fungi, insects, and erosion. The dense bark is thought to be the primary reason these trees can live fifty times longer than most other living things.

Years after this memorable hike into the bristlecone forest, Powell would remember with fondness his summer working with Schulman, the first scientist to recognize the importance of bristlecone pines for climate research. Schulman was an expert in the then-new field called dendrochronology—using tree growth rings to re-create a timeline of climate information. He had spent much of his career searching for ancient trees that could provide a means to study climates deeper and deeper into the past. His search eventually brought him to the White Mountains, where local park rangers talked about the gnarled pine trees that looked very old. After years of work, Schulman finally found and dated the oldest tree in the grove, which he named Methuselah. This tree is more than 4,800 years old—and still alive today.

Powell recalled conversations he had with Schulman around the campfire that summer. One evening in particular, the discussion took an oddly unscientific turn. Schulman talked about the trees as if they possessed the secret to longevity, and he wondered aloud about the possibility of distilling from these ancient trees an elixir of life. Not long after this summer field season, Schulman died of heart disease at age forty-nine. Although he did not discover an elixir of eternal life for himself, the trees he studied have shared their memories with scientists—memories of long droughts and periods of abundant rainfall over the millennia. By carefully linking the tree-ring patterns in dead trees with those of live ones, subsequent researchers have been able to extend the bristlecone tree-ring records back almost 10,000 years.

Trees have proven to be a powerful tool for paleoclimatologists, largely because they respond every year to conditions of temperature and precipitation, and their responses are recorded in the growth rings of their trunks. Schulman learned this in the 1930s as a protégé of one of the great tree-ring pioneers, A. E. Douglass of the Tree-Ring Laboratory at the University of Arizona in Tucson. Douglass had recognized that long-lived trees, especially those in drought-prone areas like Arizona in the Southwest, could provide important information for water planners about the long-term history of stream flow.

In the early 1940s, Schulman conducted a landmark study: a 600-year reconstruction of Colorado River flow using tree-ring records from long-lived Douglas firs. The records revealed several sustained periods of low flow, which translate into drought, that recurred with some regularity. The reconstruction showed a particularly severe drought in the late 1500s, a drought that has shown up in multiple records from throughout the West. Schulman's reconstruction also showed that the late nineteenth century and the early decades of the twentieth century were unusually wet. This was a critical period of expansion by American settlers, and climate conditions at that time gave the settlers a false impression that the region was wetter than it actually is and hence ripe for growth and the establishment of new farms.

Schulman was also interested in climate cycles—repeating patterns of relatively benign conditions with reliable moisture, followed by prolonged drought. He wanted more information about such cycles in the American West to better anticipate the region's future. Schulman did indeed find recurring phenomena, including one cycle that corresponded in length (ca. 11.6 years) to already observed sunspot activity. In later chapters, we will come back to these climate cycles, which have an important place in our understanding of climate and how it changes with time.

NATURAL REPOSITORIES OF WESTERN
CLIMATE CHANGE

Bristlecone pines are not the only—or even the most important—archives of past climates. In fact, paleoclimatologists depend on many tools and archives for their investigations. In recent decades, new technological advances have allowed Earth scientists to tease out more and more from these natural archives, ranging from the chemical and isotope compositions of fossils and sediments to molecular fossils contained within organic matter preserved in these sediments. Some are most useful for the recent past, others for events that happened thousands or even millions of years ago. All these tools have required extraordinary insights and hard work to develop, and their value continues to grow as scientists find new uses for them in their research.

Just where are these archives of past climate located, and how do we find them? Earth scientists are discovering that the modern American

West contains a wealth of evidence for past climates in some of the most unexpected places. The remarkable physiographic diversity of this region is evident in a span of just a few hundred miles. We can illustrate much of this diversity in an imaginary flight from the Pacific coast near San Francisco to the east, across California and into Nevada and Utah. In this relatively short distance, the landforms, plants, and animals change markedly, reflecting large differences in temperature, precipitation, elevation, and topography.

We begin outside the Golden Gate on the rocky coastline of the Pacific Ocean, including its magnificent kelp forests and tide pools. Beneath these dark blue and seaweed-green waters, sediments are slowly sinking to the sea floor, forming aggregates in the water column called "marine snow" that eventually settles on the seabed. The sediments, building up over the millennia, are composed of silt washed in from the land, tiny shells of plankton, and the remains of other marine life that live in the coastal waters. The shells of these tiny organisms hold information about past ambient ocean conditions, like temperature and upwelling.

Our flight moves inland over the San Francisco Bay, an estuary that receives freshwater and sediment draining nearly half the state of California. The organisms living in the bay record information about the salinity and temperature of these waters, which in turn reflect the amount of river water that flows into the bay. The shores of the bay are lined with lush tidal marshes, criss-crossed by sloughs. Continuing east but still within the sprawling estuary, the water becomes fresher and the wetland vegetation grows taller, such that a careless boater could get lost in the winding sloughs that snake through the ten-foot-tall tule bulrushes. These marshlands have been growing for thousands of years, building up layers of organic peat as the level of the sea has slowly risen. The mix of marsh plants changes over time as rainfall and runoff into the bay alter salinity, and the marsh layers contain evidence of past droughts and floods.

From the wetlands of the Sacramento Delta and its hundreds of islands, we cross the great Central Valley, where rows of irrigated crops spread out for miles on the flat valley floor as giant rectangles of green that contrast strikingly with the brown rolling hills surrounding the valley. Beneath the ploughed fields lie ancient accumulations of sediment that were eroded from the Sierra Nevada range to the east. These sediments contain layers deposited during ancient flood events.

Chaparral-covered foothills of the Sierra Nevada come into view next and, continuing upslope, forests of pine, juniper, fir, Sequoia, and cedar. The largest of these trees have been growing for hundreds to thousands of years, the thickness of their growth rings reflecting the temperature and precipitation of these past periods. Small lakes dot the landscape, accumulating sediments eroded from the surrounding watershed and windblown pollen that provides snapshots of changing vegetation over time.

Soon the majestic granite peaks of the Sierra Nevada come into view, carved and polished by glaciers during the previous ice ages into stark peaks and valleys, glistening in the sunlight. Then we make an abrupt descent down the steep eastern slope of the Sierra Nevada, where moraines—elongated accumulations of sediment eroded and transported by glaciers—were left behind as the ice age glaciers melted. These moraines mark the entrance into the drier Great Basin, with its patchwork of brown shades of earth, sparsely dotted with drought-tolerant shrubs. Here, we can see lakes encircled by what appear to be bathtub rings but are in fact ancient shorelines left in the landscape that attest to earlier, wetter periods. Finally, our transect takes us across a series of block-like mountains separated by deserts, together forming the Basin and Range Province that stretches across Nevada and into Utah. Lakes, and the remnants of ancient lakes, dot the desert landscape. The lakes have grown and shrunk over the millennia in response to dramatic climate fluctuations.

This flight over the western landscape from the coast of California to Utah has shown us many clues about the history of the region's climate. The environmental archives tell us that the prehistoric climate was not always as benign as what we have experienced in the past century and a half. They paint a picture of past droughts much larger and longer than any we have recorded in recent history, followed by enormous floods and extended wetter periods.

We will return to these archives of western climate—how they were discovered and interpreted, and what they tell us—in the next few chapters of this book. First, however, we need to sharpen our forensic skills in finding and understanding the evidence.

THE CLIMATE BACKDROP

For nearly two million years, the earth's climate has swung like a pendulum between two primary climate states: glacial (ice age) and interglacial. This period is known as the Pleistocene, and it includes the previous thirty glacial cycles. The pendulum has not kept a steady rhythm, however. The glacial periods have lasted much longer than the interglacials over the past million years. The warm interglacials are more like brief but repeating aberrations that occur just after the peak of each glacial cycle.

At the peak of the last glacial cycle, or the "Glacial Maximum," about 20,000 years ago, two-mile-thick ice sheets covered the northern United States, Canada, and northern Europe. The ice was composed of water that had evaporated off the oceans, lowering global sea level by some 360 feet. The glaciers that covered these continents are now gone, so the ice they held cannot be studied. However, detailed information about the Last Glacial Maximum, as well as the preceding cycles of ice and sea levels, are contained in the landscapes that were once buried under these ice sheets as well as in the permanent ice sheets of Greenland and Antarctica. Even sediments accumulating beneath the oceans hold critical information about past glacial and interglacial cycles.

Ice cores extracted by drilling through these high-latitude ice sheets contain records of past climates spanning hundreds of thousands of years, including information about the earth's temperature, variations in atmospheric gases like carbon dioxide and methane, solar activity, and wind speeds. Likewise, sediments have been accumulating beneath the major ocean basins for many millions of years. These sediments contain a continuous record of past ocean temperatures, sea-level change, ocean productivity, ocean circulation, and more. When compared with the ice core records, they show that, during glacial periods, sea levels dropped, and the oceans were cooler and more productive. The ice core and ocean sediment archives provide vital background information about long-term climate change on a global scale.

The Holocene period, which spans the past 11,000 years, is the most recent warm or interglacial period. It is really just a continuation of the Pleistocene era but is demarcated by Earth scientists as a separate period in recognition of the abundant information that is available about it. In the rest of this chapter, we describe the types of evidence used to piece together the climate history of the North American West over millennia.

CLIMATE CLUES

Sediments are produced as the forces of climate and water erode the earth's surface over time, yielding a treasure trove of information about past climate, including the remains of plants and animals, both the hard parts and softer organic parts. These sediments are deposited year after year, for centuries and millennia, in the low spots on Earth—such as at the bottom of lakes, bogs, marshes, estuaries, and the ocean—and these layers create what can be read as the earth's history books. Fossilized remains of plants and animals can give us clues about the prevailing environmental conditions and climate during the time the organisms were living. The nature of the sediments themselves, including the size, shape, and mineral composition of the particles, how they are layered, their color, even their odor, provide important clues that help to determine past environmental and climatic conditions. Earth scientists have used these methods for at least two centuries. More recently, over the past few decades, new methods have been developed using the chemistry and isotopic composition of the sediments to extract increasingly detailed environmental and climatic information from them.

Sediment cores have been retrieved from up and down the Pacific Coast of North America—beneath the coastal ocean, from estuaries and tidal marshes, and inland from lakes, marshes, and other settings. One of the most closely examined coastal regions in the world is the Santa Barbara Basin, just off the coast of Santa Barbara in Southern California. Researchers flock to this place to study cores of sediments that have accumulated on the sea floor over many millennia.

Initially, relatively short sediment cores, only a few meters, were taken from the basin, spanning hundreds to a couple of thousand years. These early studies showed that the sediments were laid down season after season in distinct annual layers, or "varves." This basin is ideal for preserving the bottom sediments, partly because the oxygen content of the ocean water in the bottom waters of the Santa Barbara Channel, unlike most marine settings, is extremely low. The living conditions there are inhospitable to marine life, so bottom-dwelling marine organisms cannot survive. Normally, these animals—worms, clams, and snails—churn and burrow through bottom sediments in search of food and shelter, stirring up the sediments and destroying detailed sedimentary layering.

In the Santa Barbara Basin, however, these animals are absent, and the sediment layers accumulate undisturbed, year after year, for many thousands of years. Each year, two layers of sediment are deposited: a light layer during the spring/summer, and a dark layer in the winter. Sediments that erode off the land are washed into the basin by the winter rains and form the dark layer. The light layer is composed of the remains of diatoms—microscopic ocean plants composed of silica—produced during periods of high primary productivity, or blooms, during the spring and early summer. Reminiscent

FIGURE 14. Arndt Schimmelmann and colleagues examining varved sediments cored from beneath the Santa Barbara Basin along the Southern California coast. (Photo courtesy of Arndt Schimmelmann, University of Indiana.)

of tree rings, these alternating sedimentary layers can be used to re-create a timeline by counting from the top of the core downward, with each pair of light and dark sediment layers representing one year (see figure 14).

Fossils of phytoplankton and other microscopic marine organisms contain information about changing water temperatures, coastal upwelling, nutrient levels, and ocean circulation. Pollen blown into coastal waters and preserved in accumulating sediments provides information about past vegetation from nearby coastal regions, and the composition and amount of sediments carried by rivers contain clues about precipitation and runoff, including the relative size of floods.

On land, sediments deposited in freshwater environments can also be great preservers of paleoclimate information. Lakes, ponds, swamps, and marshes—all damp, mostly low-oxygen environments—are often character-ized by slow decomposition processes that allow organic matter to be pre-served for a long time. In these sediments, the pollen, charcoal fragments, fossils, organic material, and detrital material, including their chemical and isotopic compositions, are all used in paleoclimate research. As with the coastal ocean, lake sediments are obtained by coring the bottom sediments beneath the water (see figure 15).

FIGURE 15. (*Left to right*) Frances Malamud-Roam, B. Lynn Ingram, and Christina Brady coring a small oxbow lake in the Sacramento Valley, California. (Photo taken by Anders Noren, University of Minnesota, LaCore curator.)

Lakes

Lakes are a common feature in the western landscape, and their sediments are important repositories of evidence about past climatic conditions gathered from the surrounding watershed. Particularly useful are lakes that grow and shrink with changing precipitation patterns, providing unique opportunities to assess the balance of rainfall and evaporation—a balance that changes markedly with differing climate regimes. These types of lakes, called closed basins, receive freshwater from inflowing streams and rivers and from direct rainfall, and they lose water only by evaporation from their surface.

An example of such a closed basin lake is Mono Lake to the east of the Sierra Nevada of California. Records of the size of Mono Lake over the past century and a half show that, during periods of more rainfall, the lake gradually expanded, whereas during periods of extended drought, the lake became smaller. This lake also contracted when the city of Los Angeles diverted water from its inflowing streams, starting in 1941. As the size of the lake changes, so does the elevation of its surface, measured as its lake level.

Earth scientists have devised a number of ingenious ways of assessing past lake levels in closed basins. For instance, the discovery of ancient submerged tree stumps rooted within some of these lakes has proved one of the most straightforward methods. We know that most species of trees are not able to germinate and grow with their trunks under water. Therefore, the existence of tree stumps submerged in a lake today indicates that, in the past, when the trees were living, the lake was much smaller than at present. A smaller lake implies lower precipitation and greater evaporation—thus, a dryer climate or a drought. By using the location of the tree stumps to estimate the size of the smaller ancient lake, researchers can estimate past inflow of streams relative to evaporation. In addition, by counting the annual growth rings of the submerged tree stumps, the duration of the past drought can be surmised. The death of the tree (when wetter conditions return, drowning the trees) can be determined using radiocarbon dating of the outer growth rings.

At the opposite extreme, during wetter periods, the locations of ancient lake shorelines can provide climate information. The shorelines of once-large lakes leave traces in the landscape in the form of wave-cut terraces, and these can be used to determine the size of prehistoric lakes and their surface levels. Furthermore, these terrace formations—and thus the time periods of the larger lakes—can be dated, again with radiocarbon or other radiometric techniques.

Recent technological advances have provided quantitative methods for deriving past lake levels. These include the methods used by geochemists on sediment cores taken from the bottom of lakes. Unlike tree stumps and ancient shorelines, sediment cores have the potential for providing a continuous record of changing lake size, not just the extreme highs and lows. Geochemists exploit the fact that, when a lake changes size during wetter and drier periods, its chemistry changes, and these changes are recorded in sediments or in the shells of organisms that precipitated from the lake water. During drier climates, rainfall is reduced and more water evaporates off a lake, thus concentrating the dissolved salts in the lake and causing certain minerals, such as carbonate and gypsum, to precipitate out of the lake water. These minerals then settle to the bottom, providing a layer indicative of a drought. Similarly, certain forms of elements in lake water become more concentrated during evaporation.

One such element that is commonly used by geochemists is oxygen. Geochemists measure two forms, or isotopes, of oxygen (oxygen-16 and oxygen-18) incorporated into calcium carbonate minerals that precipitate in the lake or constitute the shells of organisms that live in the lake. These minerals and shells record the relative amounts of the two oxygen isotopes (which have slightly different masses that can be measured in the laboratory on a mass spectrometer) as they form and settle to the bottom of the lake, where they accumulate and preserve this information. During dry periods, more water evaporates off the lake surface than is replenished by river inflow. Evaporation causes the lake water to become more enriched in oxygen's heavier isotope (oxygen-18). During wetter periods, rain and runoff into the lake are greater than evaporation, so the lake grows larger, and the lake water becomes more enriched in the lighter oxygen isotope (oxygen-16).

Estuaries

Estuaries, where rivers draining the land meet the sea, also contain rich records of past climatic and environmental change. For instance, the San Francisco Bay estuary has formed where the rivers draining the Sierra Nevada merge and flow together, mixing with marine waters from the Pacific Ocean. The volume of this freshwater inflow, reflecting precipitation and runoff in the bay's watershed, alter the bay's salinity. The salinity changes in turn alter the types of organisms living in the estuary—from single-celled plankton and foraminifera to invertebrate organisms like clams and mussels. When these organisms die, they settle to the bottom, their hard parts becoming entombed in the sediments, later to be brought to the surface and examined by geologists and paleontologists.

Paleontologists analyze the assemblages of fossil remains and how these have changed over time to assess changes in the estuarine environment. The geochemistry of these fossil shells also reflects the environmental conditions of the bay water, particularly the salinity and temperature. As these organisms grow and secrete a calcium carbonate shell, they incorporate elements from the water surrounding them, including oxygen. Tiny variations in the mass of these oxygen atoms can be used to provide information about past environments. River water entering San Francisco Bay through the delta is lighter than Pacific Ocean water that enters the bay through the Golden Gate, meaning it has a higher proportion of oxygen-16 compared to oxygen-18. By analyzing fossil shells separated from cores taken beneath the bay from a range of ages, a time series of past salinity can be determined, providing a long history of climate over the entire watershed draining into the estuary.

Vegetation and Plant Assemblages

Anyone who suffers from hay fever dreads the spring, when the air is full of pollen—the dust-sized gametes made by the male part of the flower. All vascular plants produce pollen in differing quantities, but not all plants disperse it in the same manner. Some plants, using showy flowers, attract insects that disperse the pollen as they fly from plant to plant, collecting food. Others produce massive quantities of pollen and release it into the wind, some of which eventually comes to rest where it can fertilize another flower. Successful as this strategy is for the plant species, the majority of the pollen never makes it to other plants.

The "leftover" pollen has proven to be a tremendous resource for scientists who are looking for clues to the past. A portion of the pollen carried by the wind eventually ends up in lakes, marshes, and even the coastal ocean, where it is incorporated into the accumulating sediments. Pollen has a naturally resistant exine, an outer wall that protects the genetic information on the inside needed for the job of fertilization. When pollen accumulates in the right depositional environments, the exine remains well preserved and recognizable for long periods of time.

For scientists, pollen grains are like fingerprints: they are unique to each plant species or genus and, ideally, identifiable by their shapes and surface markings. Equally important, the types and proportions of pollen in the sediments provide information about ancient plant assemblages and therefore about past climate conditions. As forests, grasslands, and deserts and their various combinations of plants respond to changes in climate such as temperature and precipitation, the changing mix of pollens produced paints a picture of that climate. For instance, a pollen specialist would draw very different conclusions about a past climate from sediments containing pine and spruce pollen as compared to those containing oak and sage pollen.

Wildfires

Charcoal is also found in sediments from lakes and the coastal ocean, providing information about the size and frequency of past forest fires. These fires—though sometimes catastrophic—are a natural consequence of climate, particularly in the higher elevations of the American West.

Periods of prolonged drought typically create conditions that result in larger and more frequent fires, which produce large amounts of charcoal that are carried by the wind and surface runoff into lakes, ponds, and the coastal ocean. The abundance, size, and nature of charcoal in sediment are used to assess the history of wildfires in regions throughout the West. This area of study has received much attention, with interest fueled by the growth in population and economic importance of these mountain areas. As we will see in later chapters, some of these records show a disturbing pattern of recurring large fires throughout the region.

Trees

Earlier in this chapter, we discussed how trees contain evidence of climate histories within their trunks. In addition to bristlecone pines, many other trees contain valuable information about past climate. Trees have unique attributes as paleoclimate archives: they stay in one place for centuries or even millennia, adding new growth layers, or rings, to their trunks every year. Scientists who understand the specific biological needs of particular tree species and the ecological setting of the study area use trees to "measure" ancient rainfall and temperature patterns.

Of course, no single tree contains a complete climate history. Typical tree-ring climate chronologies require dozens of individual trees in each study site, and typically several study sites are needed to produce an accurate regional climate history. To avoid cutting down large numbers of trees, researchers have devised a nondestructive method of extracting small-diameter wood cores that causes no harm. These cores are obtained by drilling through the bark from the outer edge of the trunk to the center. The extracted cores, as narrow as long pencils, are scanned into a computer, and the yearly growth rings are counted and measured. The width of the rings provides a measure of annual weather conditions. Good years (for the West, this usually means wet with a long growing season) produce wider rings, and poor years (drier and cooler) produce narrower rings. During years of extreme drought, rings may even be absent.

TELLING TIME: DATING AND AGE CONTROL

Age control is an important aspect of reconstructing past climate change. Age determinations of sediment samples from cores or other recorders of past environmental change are critical for placing the records into a larger paleoenvironmental picture and for comparing different types of records from different locations. Researchers can also use age information to calculate how quickly climate changed in the past. These past rates of change provide a context for climate changes occurring today.

Establishing a chronology for climate records is challenging, but paleoclimatologists have devised a number of techniques. The ideal method provides annual time resolution, like the annual growth rings that trees produce, which can be counted to provide highly accurate ages. Similarly, corals, mollusks, and other organisms typically secrete annual growth bands when forming their hard parts. The deposition of sediments in lakes or coastal ocean settings during certain seasons can also yield annual time resolution in the form of annual layers. As discussed above, the varves of the Santa Barbara Basin provide a good example because these layers are deposited annually, so can be counted.

More typically, though, climate records contained in sediments from the sea floor, lakes, estuaries, and marshes do not contain annual varves. Most of these sedimentary sequences require the use of radiometric dating, which is based on known rates at which certain naturally occurring radioactive isotopes decay. The familiar radiocarbon (carbon-14, also known as C-14) method is the most common approach used for samples younger than 50,000 years old. Radioactive carbon is produced in the upper atmosphere when cosmic rays directed at the earth interact with nitrogen, transforming nitrogen-14 into carbon-14. Radiocarbon, just like the more abundant (and stable) forms of carbon (carbon-12 and carbon-13), combines with oxygen to form carbon dioxide and is eventually taken up by plants during photosynthesis. Carbon-14 also dissolves into the surface waters of the ocean, where organisms use it to build their calcium carbonate shells. Once these organisms die and become incorporated into the sediment, the radiocarbon begins to decay back to nitrogen. We know the rate of decay, so we can use the initial concentration, rate of decay, and amount of radiocarbon measured in our sample to give us the age of the sample.

For sediments older than about 50,000 years, scientists can use other radiometric methods to date other components of the sediment, such as mineral grains from volcanic ash that were produced during ancient volcanic eruptions. The ash is commonly deposited in distinct, thin layers within the sediment. Another approach uses the chemical "fingerprint" of an ash layer, allowing it to be traced to a specific volcanic eruption that has been independently dated. Volcanic activity has been a common occurrence in the

American West over past millennia, so ash layers are widely used in dating lake sediments.

The precision with which a researcher can resolve time within a paleoclimate record, or "age resolution," determines the level of detail that can be reconstructed. Tree-ring records with annual resolution allow researchers to derive long histories of events that have a recurrence rate of less than a decade, such as El Niño events. At the other end of the spectrum, slowly accumulating sediments in the deep ocean may provide information only about climate shifts that occur over centuries or millennia. One of the primary factors in age resolution is the rate of growth (of a shell, for example) or accumulation (of sediment).

Sediments deposited in anoxic, partially enclosed coastal basins, such as the Santa Barbara Basin, and certain lakes accumulate at a relatively high rate of several centimeters per century. In these sediments, researchers can sample the varves and achieve annual time resolution. In the deeper parts of the oceans, far from land, sediments accumulate at a much slower rate—on average one centimeter per thousand years. These deeper ocean environments are usually higher in oxygen, and therefore some burrowing and churning by bottom-dwelling organisms searching for food means that the sediments will have been stirred, or "bioturbated," blurring the record and averaging conditions over centuries or longer. In deep ocean sediment, time resolution can be limited to thousands, or even tens of thousands, of years.

The pollen record contained in lake cores provides a sequential history of vegetation. However, because pollen accumulates very slowly, less than a quarter of a teaspoon of pollen collected for analysis may represent several decades of accumulation. Hence, the age resolution for pollen studies is normally no better than decades to centuries.

Time resolution also depends on the accuracy of dating methods. Sediments dated using radiocarbon or other radiometric techniques have age uncertainties of thirty years to upward of hundreds of years, depending on the age and environment of deposition. In some cases, age measurements can be done only at irregular intervals, in which case ages for most of the sediment core must be interpolated from just a few layers that are suitable for dating. Examples of such sporadic or opportunistic age measurements would be an ash layer that can be matched to a volcanic eruption of known age, or the first appearance of the pollen of an introduced species, such as eucalyptus.

In the next five chapters, the methods reviewed in this chapter will be essential in understanding how scientists have pieced together the history of climate over the past 20,000 years. We then turn to the more challenging questions of what causes climate to change over shorter and longer timescales and why these changes are often periodic in nature. Looking ahead to the concluding chapters, we shall also see that both accurate methods and a good understanding of past climate change will be needed to address the most difficult question of all: What will the nature of the climate of the American West be in the future, and what will it mean for the region's water resources?

SIX

From Ice to Fire

INTO THE HOLOCENE

The very words "Great Ice Age" conjure up images of a
world petrified in hundreds of thousands of years of profound
deep freeze, when our skin-clad forebears hunted mammoth,
reindeer, and other arctic animals. Deep-sea borings in the
depths of the Pacific Ocean, coral growth series from tropical—
and formerly tropical—waters, arctic and Antarctic ice cores,
and concentric growth rings from ancient tree trunks tell a
different story—of constant and dramatic swings in global
climate over the past 730,000 years. Our remote ancestors lived
through wild fluctuations from intense glacial cold to much
shorter warm interglacials that sometimes brought tropical
conditions to Europe and parts of North America. Over these
hundreds of millennia, the earth has been in climatic transition
for more than three-quarters of the time: At least nine glacial
episodes have set a seesaw pattern of slow cooling and then
extremely rapid warm-up after millennia of intense cold.

BRIAN FAGAN, *Floods, Famines, and Emperors*

ICE AGES AND THE WEST

TODAY, WE LIVE IN A RELATIVELY COOL PERIOD that began about
40 million years ago. Over this period, the earth became increasingly cool and
icy, culminating in the Quaternary Period, a dance of advancing ice sheets
(glacial periods) and retreating ice sheets (interglacials) that started two
million years ago. The most recent glacial period reached its maximum extent
about 20,000 years ago—a period dubbed the "Last Glacial Maximum." The
average global temperature then was about 18°F lower than it is today.

The earth is currently in a relatively warm interglacial period known as the
Holocene epoch that began 11,000 years ago. Whereas one view holds that
the Holocene is due to end soon, having equaled (or even exceeded) the span

of many previous interglacial periods, some scientists believe that it could last thousands of years longer, perhaps extended by the addition of heat-trapping carbon dioxide and methane into the atmosphere when humans began large-scale agriculture and, most recently, the burning of fossil fuels. In truth, no one knows precisely when the next ice age will return.

The Quaternary has been divided into more than twenty glacial periods, separated by interglacial periods. The different glacial states are marked by changes in oceanic currents, changes in winds and airborne dust, and shifts in storm tracks influencing when and where precipitation falls. Both temperature and carbon dioxide levels have fluctuated between glacial and interglacial states (see figure 16A), based on ice core records discussed in the previous chapter. During glacial periods, sea levels have fallen by up to 120 meters (390 feet) below its present, interglacial level as huge volumes of water were transferred from the oceans to the continents to be stored as ice in massive ice sheets. The lowered sea levels exposed a land bridge between Asia and North America.

When the last glacial period of the Quaternary reached its peak about 20,000 years ago, one of its ice sheets—the Laurentide—was two miles thick and covered most of what is, today, eastern and central Canada, extending southward into Indiana, Illinois, Ohio, and Pennsylvania. On the western half of Canada was the Cordilleran ice sheet that reached from British Columbia as far south as Seattle, Washington. These ice sheets grew so high that scientists believe they altered atmospheric circulation. Mountain glaciers also expanded over the upper reaches of the Cascade, Rocky, and Sierra Nevada ranges.

Our story focuses on the climate history of the American West since the Last Glacial Maximum. Within this relatively short time period, climatic extremes were greater than anything we can imagine from our limited historical experience. Today we live in a time of unusually benign weather, and many "normal" conditions, as viewed over the past 20,000 years, seem inconceivably harsh.

After the peak of the last ice age, the climate began to warm at last, and the ice began to melt, carving new rivers and leaving behind mountain-ous loads of debris in moraines. Fifteen thousand years ago, enormous lakes formed across the Great Basin of modern-day California, Nevada, Utah, Oregon, and Colorado. Some of these lakes were tens of thousands of miles in area and hundreds of feet deep, covering vast basins between the mountains.

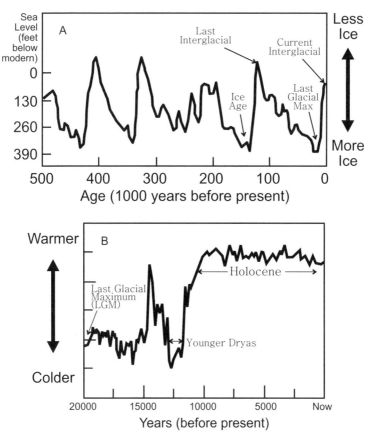

FIGURE 16. (A) Sea level changes (in feet below modern sea level) over the past 500,000 years, showing glacial/interglacial cycles. During glaciations, sea level dropped as seawater was transferred to the poles to build huge ice sheets; during interglacials, the ice sheets melted to release the water back to the oceans, raising their level by 390 feet. (B) Temperature over the past 20,000 years measured in the Greenland ice cores, showing the transition from the Last Glacial Maximum, 20,000 years ago, to the Holocene epoch, which began 11,000 years ago. The climate returned to near glacial conditions during a 1,300-year period called the Younger Dryas. (Drawn by B. Lynn Ingram based on open online sources.)

The first human immigrants made their way to North America from Asia across a land bridge between Siberia and Alaska that would later be swallowed up by the rising sea. They likely entered after the planet warmed sufficiently to melt a narrow corridor between the two enormous ice sheets over Canada, some 13,500 years ago. Archaeologist Brian Fagan reasons that a southward migration through this long corridor (approximately 930 miles) may have taken generations, probably as people followed the seasonal migrations of animals such as caribou and bison.

Upon their arrival in the new world, people would have found plenty of food, as well as predators such as saber-toothed tigers, and rapidly changing habitats. Around 10,300 years ago, the gradual warming trend that had marked the previous several thousand years in the West suddenly ceased and the climate deteriorated. For the next several hundred years, glacial conditions returned, a period known as the Younger Dryas. As suddenly as that period started, it ended, and the warming resumed, marking the beginning of the Holocene (see figure 16B). Melting glaciers once again sent water rushing into the sea, raising the level of the oceans worldwide by nearly an inch a year. The rate of sea level rise was quite high until about 6,000 years ago; by then, the rising seas had flooded what had been coastal river valleys, forming the bays, estuaries, and wetlands that are part of our modern coastline.

VAST ICE AGE LAKES IN THE GREAT BASIN

Climate scientists have learned a great deal about deglaciation (a period of rapid climatic warming and melting ice sheets) by studying enormous lakes in the Great Basin. The Great Basin is a region in the heart of the American West that extends across the states of Nevada, Utah, Oregon, and into Colorado, bracketed on the west by the Sierra Nevada and Cascade ranges, on the north by the Wallowa Mountains, and on the east by the Rocky Mountains. The region comprises a series of linear mountain ranges and basins that formed over millions of years of tectonic movement between the Pacific and North American plates, causing extensional pulling and shearing of the landscape. Today, it is predominantly a high desert region, beautiful and stark with sagebrush-dominated desert valleys alternating with linear mountain ranges. The Great Basin is so named because of its hydrology: the region's rivers have no outlet to the ocean, instead terminating in the lakes, vast playas, and dry lakebeds of this giant basin.

Climate in this region today is semiarid, with mild winters, warm summers, and little precipitation. Like the West Coast, precipitation is brought to the Great Basin mostly during the winter by westerly winds carrying moisture from the Pacific Ocean. Rain and snow from the mountain peaks feed the rivers that drain into the Great Basin, but freshwater is limited, not in sufficient supply for its growing population.

The Great Basin valley floors contain huge expanses of salty sediments that are the remains of vast ice age lakes. The Great Salt Lake desert, covering some 20,000 square miles of western Utah, is one such region. It was made famous in August 1846, when the Donner pioneer family joined the Reeds and several other families migrating to the West on what was supposed to be a shortcut across this desert. Instead, they encountered a series of setbacks, including a lack of freshwater and thick salty clay sediments that proved heavy going for their ox-drawn wagons, all of which took a toll. The result of using this shortcut (the so-called Hastings Cutoff) was a delay that put the migrants in the Sierra Nevada in October, exactly the wrong time to begin a crossing of this mountain range. The timing was made worse by the onset of an early winter with a series of heavy snowstorms. The misery that ensued is now an ingrained part of California's history.

If we step further back in time and observe this same region 15,000 years ago, the desert was instead part of a vast, deep body of water, which is today called Lake Bonneville. Had the early pioneers crossed the region during that time, they would not have been contending with empty, endless deserts; rather, they would have been following seemingly endless shorelines that bordered gargantuan lakes. Today, the Great Basin has some sizeable lakes and playas, including Pyramid Lake in Nevada, Owens Lake in California, and the Great Salt Lake in Utah. Drawing on a variety of lines of evidence, both from within the lake-bed sediments and from the geomorphic traces of ancient shorelines that ring the playas along the slopes of nearby mountains, it is clear that, at times in the past, the Great Basin supported vast lakes that were not only hundreds to thousands of square miles in area (for example, the Great Salt Lake's ancestor, Lake Bonneville, was close to 30,000 square miles at its peak) but also very deep. At one time, Lake Lahontan, in northeastern Nevada, was one of the largest lakes in North America. With over 10,000 square miles in area, this lake once encompassed modern-day Walker Lake in the south, Pyramid Lake on its western edge, and Winnemucca Lake on the northeastern edge. Yet another large lake, Owens, is found in southeast California. Today, this lake is completely dry, but it once covered an area close to 400 square miles. These lakes expanded and contracted through the ages with changing climate.

Enormous lakes were, in fact, a common feature on the glacial landscape. But, in contrast to the colossal glacial lakes that formed farther north along the southern edges of the melting continental ice sheet, it is thought that the lakes of the Great Basin grew to such a great size as the result of both higher

precipitation in the West during the ice age and cooler air temperatures, causing less evaporation off their surfaces. During the Pleistocene, eight lakes in
the Great Basin covered some 27,800,000 acres, eleven times the surface area
covered by lakes today.

One of the largest of these Pleistocene lakes was Lake Bonneville. Mapped
by G. K. Gilbert in the late nineteenth century, Lake Bonneville once reached
a size equal to Lake Michigan, covering the northwest corner of modern-day
Utah and extending into Nevada and Idaho (see figure 18 in the next chapter). At over a thousand feet deep, it was significantly deeper than modern
Lake Michigan. The weight of this lake was so great that it depressed the
underlying land surface some 240 feet, and, when the lake later went away,
the land surface rebounded. Gilbert mapped flood features just to the north
of Lake Bonneville and hypothesized that the lake may have burst its dam
catastrophically about 15,000 years ago, creating a megaflood that flowed
into the Snake River Basin in southern Idaho and finally into the Columbia
River. Today, the only remnant of this enormous ice age lake is the Great Salt
Lake in Utah, so named because it is one of the saltiest lakes on Earth—seven
times saltier than the ocean.

Glacial summers were cooler in the northern hemisphere than they are
today, largely as the result of astronomical factors that conspired to reduce,
ever so slightly, the total amount of incoming solar radiation and to mitigate the extremes of the seasons. These cooler summers allowed the enormous continental ice sheets, including those that covered Canada and the
northernmost United States, to grow to two miles high—a height that was
sufficient to influence atmospheric circulation patterns.

Climate scientists have run computer models showing that the jet
stream, a high-speed wind current that brings winter storms to the Pacific
Northwest and as far south as Northern California today, may have
been split into two limbs as it encountered the ice sheet. These models suggest
that the southern limb of the jet stream during the late Pleistocene would
have brought Pacific storms into the Great Basin, increasing precipitation
during the winter. Cooler, cloudier conditions during the summer would
have also decreased evaporation in the Great Basin, where today an average of
15 to 30 inches of water evaporates off the lakes every year. The overall effect
would have been an increase in moisture in the region—and the expansion
of lakes.

Paleoclimatologist Larry Benson and his colleagues at the U.S. Geological
Survey have spent their careers studying Great Basin lakes, noting that those

in the northern Great Basin—Lahontan and Bonneville lakes—were about ten times larger during the late Pleistocene than today. Searles and Russell lakes in the southern Great Basin, in contrast, expanded by a factor of only four to six. Benson hypothesized that, during the late Pleistocene, the position of the jet stream may have been predominantly situated over the northern Great Basin, bringing many more rain-bearing storms to that region.

THE THAWING OF THE WEST

The world that the earliest human immigrants would have found was quite different from the world we know today. Large regions of North America were still shrouded by ice, and glaciers blanketed the mountain ranges, reaching down to much lower elevations than during the warmer Holocene.

As the mountain glaciers flowed downslope, they carved out valleys. Later, when the ice had retreated, these valleys would be left with a characteristic "U shape," with steep sides and wide, flat valley floors. When the first human immigrants were crossing the land bridge (over what is called the Bering Strait today), the floors of these glacial valleys, like Yosemite in California's Sierra Nevada, were still buried under hundreds of feet of snow and ice, with only the highest peaks standing above the ice. Cold temperatures during this glacial period displaced the forests in the mountains to lower elevations and lower latitudes. The plant assemblages in the mountains differed as well: firs and certain pines were the dominant trees, and sagebrush communities tolerant of cold, dry conditions were more common at the lower elevations.

After the peak of the last ice age, when the world began to thaw, conditions in the western mountain ranges changed rapidly as the alpine glaciers melted, leaving behind accumulations of rock, gravel, and sand that had been scraped off the canyon bottoms and walls. These deposits were massive enough to block streams swollen with glacial meltwater, forming vast glacial lakes, hundreds of feet deep. Yosemite Lake, for instance, was formed when the ancient Merced River was dammed, forming a lake that covered a region over five miles long. Half Dome and El Capitan towering above the lake would have been reflected in its surface. The lake bottom now forms the flat valley floor of Yosemite Valley. In the higher Sierra Nevada, lakes formed by glacial dams still exist today.

The Great Basin lakes reached their peak in extent about 15,000 years ago. Lake Lahontan reached a high stand during the late Pleistocene, 15,500 years ago, then began to shrink, reaching an area of only one-seventh the size by the early Holocene, 10,000 years ago. Other lakes showed similar declines, such as the Great Basin's Owens Lake in California and Lake Bonneville in Utah, and, outside the Great Basin, Tulare Lake in California's Central Valley, all of which reached low stands coinciding with the end of the Pleistocene and the start of the Holocene.

CLIMATE AND THE FIRST WESTERNERS

Humans arrived in the western landscape around 13,500 years ago. These stone-age hunter-gatherers, the so-called Clovis people who left behind pointed spears fashioned from stone, migrated south from Alaska into the Plains and Great Basin and continued into the Southwest and California. They settled close to sources of reliable water: lakes, springs, and perennial streams. When the climate deteriorated and water sources dried up, they moved on. Their low population numbers gave them flexibility and mobility, which were key to their survival. From the coast to inland valleys, these earliest westerners could choose from a wide diversity of foods: fish, including salmon during their spring and fall migrations upstream to spawn; small and large game; wild plants; and migratory birds that stopped over in wetlands like California's Central Valley en route to more southerly or northerly feeding and breeding grounds. Strange ice age animals—the Pleistocene "megafauna"—also roamed the region, including giant sloths, herds of mastodons and mammoths, native camels, horses, and elk, as well as fearsome predators like the saber-toothed tiger and the grizzly bear.

Along the coast, food would have been plentiful—particularly mollusks, fish, and sea mammals—but the early inhabitants would have had to contend with rapidly changing environments during the late Pleistocene and early Holocene. During this time of transition, the coast was perhaps one of the most rapidly changing environments in the West, its configuration changing dramatically with the meltwaters of glacial ice flowing into the oceans. Between about 18,000 and 6,000 years ago, sea level rose an average of three-quarters of an inch per year, transforming coastal river valleys into shallow bays and estuaries.

Just before the Pleistocene ended, about 11,000 years ago, the steady shrinking of lakes in the Great Basin briefly halted and reversed for several hundred years. Lake Bonneville expanded during this period, creating the so-called "Gilbert Shoreline." This short-lived return to greater moisture in the American West coincided with a brief advance of glaciers in Europe, during what has come to be known as the Younger Dryas event (see figure 16B). Evidence from Greenland ice cores suggests it lasted about 1,300 years in the North Atlantic region, though its duration appears significantly shorter in the American West. In the broad timescales of climate, this return to glacial conditions was so brief that paleoclimatologists think of it as an "event," though one that evidence suggests may have been hemispheric, if not global.

One hypothesis explaining the cause of the Younger Dryas relates to another great Pleistocene lake. At its peak, the area of Lake Agassiz was larger than all the Great Lakes combined. Named after Louis Agassiz—the father of glacial geology—this lake owed its massive size to an ice dam that backed up the glacial meltwater as the Pleistocene ice age was ending. This enormous volume of water was abruptly and catastrophically released after the ice dam collapsed, with the resulting wave flowing into Hudson Bay, through the St. Lawrence Seaway, and into the North Atlantic, a critical location for deep ocean circulation. In this region, salty Gulf Stream waters, which have been transported northward, are cooled, becoming dense enough to sink to great depths, forming deepwater that flows slowly southward. But when the rush of fresh lake waters burst out into the North Atlantic, they would have floated on the surface (freshwater being less dense than saltwater, therefore more buoyant), preventing the sinking of cold, salty water that forms the deepwater. This, in turn, slowed the normal northward flow of warm, salty Gulf Stream water to the region.

Today, this warm current heats the atmosphere above it along the coastlines of the continents surrounding the North Atlantic. When the current was slowed, or even stopped altogether, the result would have been a dramatic cooling. These changes in atmospheric and oceanic circulation that occurred during the Younger Dryas appear to have affected regions as far away as California and the American West, where the Great Basin lakes once again grew large.

The Younger Dryas as well as other major global climate changes influenced conditions along the Pacific Coast. Evidence for these impacts is found in sediment cores from the eastern Pacific Ocean, including those accumulating in the Santa Barbara Basin off the Southern California coast. As described in the preceding chapter, sediments are laid down there, year after year, leaving behind distinct layers. During warm interglacial periods, the deepwater entering the basin bottom is very old and depleted in oxygen, so that bottom-dwelling marine animals cannot survive to mix and churn the sediments. The basin sediments form in undisturbed seasonal layers, with light-colored layers resulting from the siliceous diatoms raining down the water column during the spring and summer, and darker layers resulting from the silts and clays transported with runoff from the land during the winter. But was this also the case during the ice ages and the Younger Dryas event?

This was one of the questions that James Kennett, a founding father of the field of paleoceanography—the study of ocean and climate conditions of the past—set out to answer. Kennett was the co–chief scientist for the Ocean Drilling Program's coring efforts in the Santa Barbara Basin in the early 1990s. There, he and his team retrieved the longest cores from the eastern Pacific to date, spanning the past 160,000 years. These cores reveal that, during the last ice age, Santa Barbara Basin sediments lacked their unique annual layers. In a collaborative effort between Kennett and one of this book's authors, B. Lynn Ingram, we discovered that, during the last ice age, ocean circulation appeared to have been altered, with the flow of younger, more oxygenated waters to the eastern Pacific Ocean and into the basin. These more oxygenated waters would have supported bottom-dwelling organisms—worms, clams, and so on—that churn through the bottom sediments as they dig burrows and search for food, destroying the fine laminations. Kennett had observed that the dark brown and finely layered sediments typical of the late Pleistocene and Holocene sequences in the Santa Barbara Basin were abruptly replaced by a three-foot-thick layer of light-colored, homogeneous sediment during the Younger Dryas, a period when younger, more oxygenated waters entered the basin.

Kennett separated out the fossil shells of tiny, single-celled protists called foraminifera (or "forams") from the sediments to extract from their chemical composition information about the water conditions in which these organisms grew. From these shells, Kennett has been able to reconstruct the temperature of surface waters surrounding the forams using the relative proportions of the heavier isotope of oxygen (oxygen-18) compared to the lighter isotope (oxygen-16). The oxygen isotopic ratios (oxygen-18/oxygen-16) preserved in the calcium carbonate shells of these forams showed that coastal sea surface temperatures dropped by 3°C (5.4°F) during the Younger Dryas event.

Kennett's research team hypothesized that this homogeneous sediment layer was formed when younger, oxygen-rich waters flowed into the basin to replace the older, oxygen-depleted bottom waters. But how did these oxygen-rich waters suddenly appear in the basin and why only during this geologically brief period? Just as during the ice age, the oceanic and atmospheric changes during the Younger Dryas may have caused an increase in deep and intermediate water formation in the North Pacific. These waters would have reached the Santa Barbara Basin more quickly, with less time for the oxygen to be removed by bacterial decay of organic matter. Our radiocarbon analyses of the foraminifera support this hypothesis, showing us that major changes in the circulation patterns of the ocean and atmosphere can occur quite rapidly. Studies of ice cores reveal that such changes can occur within decades.

Evidence of this major climatic cooling during the Younger Dryas has not been found in all paleoclimatic records throughout the American West. In some regions, only subtle impacts were found, with small shifts in vegetation—such as increases in pine and decreases in alder and oak in some of the mountain areas. Furthermore, as we saw above, many of the Great Basin lakes expanded, as did California's Tulare Lake. Some scientists suggest (though not all agree) that the West's climate briefly returned to colder and wetter conditions during the Younger Dryas.

THE GREAT WARMING: INTO THE HOLOCENE

The Younger Dryas ended almost as abruptly as it began, ushering in the Holocene, our current period of climatic warmth. As the climate began to warm and dry, the lakes again shrank. The seasons were also more pronounced at that time in the West, because the amount of solar radiation hitting the Northern Hemisphere during the summer was about 8 percent higher than today, so summers were warmer and winters were cooler. The increased summer temperatures were caused by slight changes in the earth's orbit around the sun that changed the amount of solar radiation received by the earth. In the early Holocene, the earth's axis was tilted slightly more toward the sun, and, during the Northern Hemisphere summer, the earth was slightly closer to the sun, resulting in an increase in solar radiation during that season. These changes in the earth's orbit over time are known as the Milankovitch Cycles. We describe these cycles and their influence on the earth's climate in more depth in chapter 11.

Pollen evidence of vegetation change suggests that the transition from the late Pleistocene to the early Holocene brought warming and drying to the American West, concurring with the Great Basin lake-level evidence. Just as the massive lakes of the Great Basin were shrinking, Sierra lakes were forming in California: the massive alpine glaciers had scoured out the bedrock, and their moraines later dammed these carved valleys, forming chains of mountain lakes. Freshwater, supplied by glaciers melting in the higher elevations, fed the mountain lakes even as overall conditions were drying.

Forests usually colonize the mountainsides in the wake of retreating glaciers. Natural vegetation zones exist along the steep mountain slopes, tied to elevation and climate. In the higher elevations within the forested slopes, pine trees dominate, even throughout major climate shifts. Paleoclimate researchers look for particularly sensitive locations at mid and low elevations, where meadows, marshes, and lakes are found. These sensitive environments contain sedimentary archives that record the regional environmental history. Relatively small changes in climate—such as a slight warming or a change in effective moisture availability—affect the local plant assemblages as the vegetation shifts in response to the new conditions. The forests surrounding small mountain lakes or meadows produce pollen and other macrofossils that are blown or washed into the lakes, to be incorporated into the slowly accumulating sediments on the lake bottom. Researchers can derive a series of snapshots from the sediments that reflect changes in the local flora over time.

One such lake, called Lake Moran, has provided a wealth of paleoclimatic information about the transition from the late Pleistocene to the Holocene. Moran is a small lake set in a mid-elevation forest, a bit over 6,600 feet, on the western slopes of the Sierra Nevada. It is the type of setting desired by paleoclimate researchers: set well into the forest away from any main roads, the lake is like a jewel surrounded by sugar pines, ponderosa pines, and fir trees. The lake itself drains a relatively small area (about thirty acres) with no inlet stream, so that climate researchers infer that the pollen, charcoal, and other fossil evidence found in the sediments are derived locally. Since the late 1980s, Roger Byrne, a paleoclimatologist and pollen expert in the Geography Department at the University of California, Berkeley, has used fossilized pollen and other macrofossil evidence to decipher past changes in climate around Lake Moran. Together with Eric Edlund, then a graduate student, Byrne collected several cores from the lake.

Radiocarbon dating of these cores showed that Lake Moran formed 15,000 years ago, after glacial ice began retreating in the Sierra Nevada. The sediment cores were also dated by matching layers of volcanic ash that settled into the lake with eruptions from nearby Mono Craters. An ash layer provides a basal date of the lake as 14,500 to 15,000 years old. Over the millennia, the lake has been slowly filling with sediments from its small watershed. These sediments contain a record of environmental change that spans the critical transition from the late Pleistocene into the Holocene.

Lake Moran contains seasonal layers: dark, organic rich layers deposited during the winter, and lighter, denser layers during the summer. In total, about ten feet of sediment have accumulated in the lake during its 15,000-year history, each foot of sediment containing about 1,500 years of environmental and climatic history.

The record of environmental change that Byrne and Edlund reconstructed from these sediments contains insights about how the region experienced the climatic transition at the end of the relatively moist and cool Pleistocene and the subsequent climate fluctuations of the Holocene. They found some surprises in this record. The notion so commonly held by ecologists of "plant communities," with specific species living in close association in specific habitats, appears to be a function of particular climatic conditions. Edlund and Byrne found some unlikely combinations of plants growing in close proximity to each other around Lake Moran at times in the past that would never be found today. For example, about 12,000 years ago, a dense coniferous forest surrounded the lake for a period lasting between one thousand and two thousand years. Sugar pine, which prefers warm summers, grew with mountain hemlock, which prefers wetter conditions. Today, the sugar pine is generally restricted to the lower elevation forests in the Sierra, and the mountain hemlock is found at higher, moister locations. The fact that these were found together suggests a very different climate regime in the Sierra in latest Pleistocene, one with warm summers but abundant moisture. These plant assemblages have no analog in modern vegetation communities.

We can infer certain things about regional climate conditions from these unusual combinations of plants. Although the region was generally warming rapidly, the spring season was apparently cooler in the Sierra Nevada than today, with more snowstorms than rain, and the snowpack persisted later into the summers. Edlund and Byrne concluded that the Sierras must have lacked the pronounced summer drought that is so characteristic of the region today. The pollen preserved in Lake Moran sediments also shows that the early Holocene was a time of rapid change. Over the course of several thousand years, the pine forests thinned, transforming the region into a more open landscape occupied by drought-adapted oaks as the climate continued to warm and dry. This warming reached a maximum about 7,000 years ago.

Pollen evidence from other lakes throughout the West also suggests a period of warm and dry climate during the early Holocene. Closed-basin lakes grew smaller as the climate grew drier. These lakes include Pyramid Lake in western Nevada and Owens Lake in eastern California, both of which decreased in size between 8,000 and 6,500 years ago.

CHARCOAL AND WILDFIRES

The climatic trends indicated by vegetation change are also reflected in the frequency and intensity of wildfires in the West. For instance, the abundance and size of charcoal fragments that reached Lake Moran increased during times of large forest fires. These charcoal fragments settled to the bottom of the lake and became part of the bottom sediments. Charcoal can range in size from fine dust particles to large chunks up to an inch or more across. The finer particles are carried longer distances in the wind. Furthermore, larger, more intense fires produce higher plumes of air that rise above them, carrying charcoal farther from the site of a fire. Because large forest fires often occur during periods of warmer and drier climate (particularly if they follow a wetter period, which increases the amount of vegetation that later becomes fuel), the size and amount of charcoal fragments in buried sediments provide additional evidence for past climate conditions.

At Lake Moran, the charcoal record suggests that large fires were uncommon during the late Pleistocene. This comes as no surprise, given the wetter and cooler climate that prevailed. But during the transition into the early Holocene, charcoal abundances increased as summers became warmer and drier. This increased amount of charcoal from forest fires remained high until about 7,000 years ago. In fact, the early Holocene appears to have been a period of more extensive and intense forest fires than any time since. Byrne and Edlund speculated that the large number of forest fires may have significantly shaped forest communities, and therefore fire would have been the means by which the early Holocene climate change dramatically shifted the vegetation from a dense forest regime to a more open forest dominated by oak trees.

HUMANS AND FIRE

Although fires and climate largely controlled vegetation shifts during the Holocene, some archaeologists argue that early humans may have also played an important role in shaping the types of plant communities in some parts of the West. These researchers suggest that native populations managed the land over the millennia with controlled burning, pruning, weeding, irrigation, and tillage. In other words, the early humans were not merely passive hunter-gatherers who collected their food from a wild landscape, as was once thought.

According to Kent Lightfoot and Otis Parrish at the University of California, Berkeley, hunter-gatherer populations may have used controlled burning to clear underbrush and areas of dead timber, eradicate pests and diseased plants, and recycle nutrients back into the soil. Fires may also have been used to clear water holes of vegetation, drive wildlife into traps during group hunting, and mark territories or claim resources. The geographic extent and intensity of these controlled burns over time is a subject of Lightfoot's current research, and, in his work with Parrish, he has hypothesized that some of the landscape patterns we see over the past 13,500 years or so after humans arrived in the West were related to intentionally set fires, not just those ignited naturally by lightning strikes.

EARLY HOLOCENE CLIMATE CYCLES

Broad cycles of climate change during the Holocene have been apparent from the early part of the epoch. The evidence is found both on land and in the oceans. The bristlecone tree line migrated downslope as climate cooled and upslope again as climate warmed, approximately every 2,500 years. These broad cycles in temperature are also apparent in the waxing and waning of mountain glaciers.

Off the West Coast in the Pacific, there is also evidence that seawater temperatures shifted with similar broad cycles—temperature shifts that affected the assemblages and oxygen isotopic compositions of coastal marine organisms. James Kennett and his research group analyzed the fossil remains from sediment cores taken from the Santa Barbara Basin. They used two species of foraminifera: *Globigerina bulloides,* a surface dweller, and *Neogloboquadrina pachyderma,* a species that lives at about 60 meters' (195 feet) depth, where ocean temperatures become much cooler. Their results show that coastal waters were unusually warm (by 2–3°C, or 3.5–5.5°F) from 11,000 to 10,300 years ago. After that time, during the earliest Holocene, temperatures began to drop, cooling by 4–5°C (7–9°F), and remained low for the next 1,500 years. Temperatures then increased again between 8,200 and 6,700 years ago, and they fell once more between 6,700 and 5,200 years ago.

These millennial-scale climate cycles during the Holocene were initially identified about a decade ago in climate records from Greenland ice cores and North Atlantic sediment cores. Since then, researchers have found evidence for them in pollen records showing vegetation change throughout North America as well as marine records from the Pacific Ocean. We will return to these broad cycles of climate in the next few chapters and will discuss their possible causes in chapter 11.

The transition from the peak of the last ice age to the Holocene encompassed some of the most rapid and monumental climate shifts in the geologic record. Global cooling at the peak of the last ice age is important because it represents a period when the climate was quite different from the warm conditions we experience today and are likely to experience in the future. The evidence shows that the West was significantly wetter during this cooler period, resulting in the formation of the enormous Great Basin lakes, which were ten times larger than they are today. These lakes have since largely dried up, and they are predicted to dry even more as the climate continues to warm. As these lakes gradually desiccate into salt flats, they serve as a warning of how rapidly water can vanish in a region that is highly sensitive to climate change.

Compared to the monumental environmental and climatic upheaval of the previous ice age, the Holocene epoch was a more tranquil period. There were no wild expansions and contractions of great lakes; no large, abrupt shifts in temperature. Even so, the climate fluctuations and extremes of the past 11,000 years have been large enough to dramatically alter the landscape, disrupt human societies, and, in some cases, lead to mass migration and societal collapse. As we shall explore in the next chapter, the most extreme of these fluctuations was one of the longest and most severe droughts of the Holocene in the West.

SEVEN

The "Long Drought" of the Mid-Holocene

> There are only a limited number of ways societies can respond
> to accumulated climatic stress: movement or social collabora-
> tion; muddling their way from crisis to crisis; decisive, central-
> ized leadership on the part of a few individuals; or developing
> innovations that increase the carrying capacity of the land. The
> alternative to all these options is collapse.
>
> BRIAN FAGAN, *Floods, Famines, and Emperors*

DURING THE MID-HOLOCENE, CLIMATE IN THE WEST shifted toward warmer, drier conditions. This climatic upheaval and prolonged drought forced many humans to leave their home terrain in search of water and food, especially in the Great Basin and what is known today as southeastern California. The inland areas were the hardest hit, and, in response, the archaeological evidence suggests that populations migrated to the coast, which offered a cooler climate, more moisture, and plentiful food.

Along the coast, the steady rise of sea level following the last ice age slowed after the massive inflow of glacial melt waters began to taper off. Coastal habitats began to expand, and coastal ecosystems flourished. Coastal groups took advantage of these abundant marine resources, establishing settlements along the Pacific Northwest coast, around present-day San Francisco Bay, and as far south as California's Santa Barbara Basin and Channel Islands.

Coastal conditions also became more favorable for salmon migration, spawning, and population growth. Salmon soon became the single most important Native American food resource along the coast, attracting still more humans to the area.

TAHOE'S SUBMERGED TREES

Some of the most vivid evidence of the mid-Holocene drought comes from Lake Tahoe, the largest alpine lake in the United States. Appropriately

FIGURE 17. Ancient tree stumps exposed in southern Lake Tahoe in the Sierra Nevada, California. (Photo courtesy of Derrick Kelly.)

enough, evidence of the prolonged mid-Holocene drought was discovered during a modern drought—the Great Dust Bowl drought of the twentieth century, which was California's worst multiyear drought in recent times. Four years after that drought began in California in 1928, the water level of Lake Tahoe in the Sierra Nevada had dropped by fourteen inches, exposing a mysterious clustering of tree stumps sticking up from the water's surface along the lake's southern shore (see figure 17). These trees attracted the attention of Samuel Harding, a scientist at the University of California, Berkeley, who discovered that the trees were large, with trunks upwards of three feet in diameter, and that they appeared to be firmly rooted in the lake bottom.

Wanting to know more about these trees, Harding collected eleven core samples from their trunks. He reasoned that the trees had grown in this location for a long time to attain such sizes and, since they were now submerged in over twelve feet of water, that at some time in the past the lake level had been much lower. He counted 150 rings in one of the cores, which indicated that south Lake Tahoe had been dry and below the shoreline for well over a century. The site must then have been abruptly inundated with water, drowning the trees fast enough that they were left standing in their current upright positions.

Harding considered several possible explanations for the location (and elevation) of these trees. A large tectonic event might have lowered the southern shoreline where the trees were found or raised the outlet of the lake. Alternatively, an extreme climate event involving an extended dry spell might have caused the lake level to drop, allowing the trees to grow along the former shoreline. If the drought was followed by a return to wetter conditions, the lake would have risen, submerging and drowning the trees. The tree stumps were minimally decomposed, suggesting that, whatever the cause, the drowning had been rapid.

Harding found little evidence to support a sudden tectonic event; no visible signs of this could be seen in the sediments or landscape. Moreover, submerged tree stumps had been found in other nearby lakes, suggesting a more regional cause and leading Harding to ultimately favor the drought hypothesis. But, in the mid-1930s, he did not yet have the tools to date the trees, and he was forced to put aside the mystery of these tree stumps for a later time. It would be three decades before he was able to piece together the story.

In the years following Harding's original discoveries, scientists found more tree stumps along the southern shore of Lake Tahoe, submerged at depths ranging from two to twelve feet below the modern water surface. Since Lake Tahoe is the largest and second-deepest alpine lake in North America, only a very prolonged and severe drought could have lowered its surface level enough to allow trees to establish and survive for centuries. But when did this devastating drought occur? Fortunately, a method of dating fossils and artifacts containing carbon (organic carbon or carbonate) by measuring their radioactive content was invented in the 1950s. Using this new technique, Harding himself had the gratification of finally dating the outer-most rings of the tree stumps to determine the time of their death. Sampling three of them, he found that the trees had been flooded and killed approximately 4,800 years ago, in the mid-Holocene.

RE-EMERGING TREE STUMPS

The late 1980s were marked by another historic drought: California's second most severe drought on record. Lake Tahoe's surface level dropped again to expose more tree stumps. This time, it was an archaeologist, Susan Lindstrom, who noticed the tops of trees sticking out of the water along Tahoe's southern shore. Donning scuba gear, Lindstrom was able to find fifteen submerged tree stumps that had escaped Harding's attention, some measuring up to three and a half feet in diameter and seven feet tall. The core samples and radiocarbon dates from this much larger population of trees refined and extended the boundaries of the mid-Holocene drought, moving its beginning to approximately 6,290 years ago and its ending to 4,840 years ago. These stumps, located deeper in the lake, showed that the lake level had dropped by even more than Harding originally thought—by more than twenty feet. Lindstrom has since located tree stumps in more places around the shores of Lake Tahoe, including Emerald Bay, Rubicon Point, and Baldwin Beach. Stumps of similar age have also been discovered in other nearby Sierra lakes, including Donner, Independence, and Fallen Leaf.

The lowered shoreline of Lake Tahoe in the late 1980s revealed other interesting traces of the past. Archaeologists discovered bedrock mortars containing grinding cups and milling slicks on the west shore, close to Tahoe City, providing evidence that humans lived there during the mid-Holocene. These artifacts and others from archaeological sites around the Tahoe Basin confirm that humans were able to survive in the region despite the prolonged drought.

PYRAMID LAKE

More evidence of the mid-Holocene drought has been found in Pyramid Lake, located in northwestern Nevada. This lake is hydrologically connected to Lake Tahoe by the Truckee River, the only outlet for Lake Tahoe. After leaving Lake Tahoe, the river flows down the steep eastern flank of the Sierra Nevada, through Reno, and forty miles northeast into Pyramid Lake. Under natural conditions, therefore, the size of Pyramid Lake is determined in large part by the level of Lake Tahoe.

Pyramid Lake is a remnant of the ancient Lake Lahontan, which occupied a large part of northern Nevada at the end of the last ice age (see figure 18). Pyramid is a relatively deep lake, at 890 feet, and has no outlet. This makes it ideal for paleoclimate research. Since 1970, Larry Benson, a geochemist and hydrologist at the U.S. Geological Survey, has been studying what the lake can tell us about the West's climatic history, including the relationship between past climate change and human culture.

One of Benson's achievements has been the development of methods for assessing past lake levels by measuring the oxygen isotopes in calcium

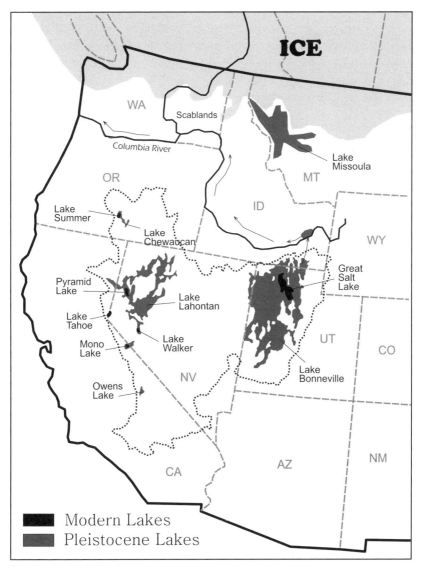

FIGURE 18. Map of the distribution of enormous late Pleistocene lakes, including Lake Bonneville in Utah and Lake Lahontan in western Nevada. Modern lakes discussed in the text are also shown, as well as the Cordilleran ice sheet over western Canada that extends into northern Washington, Idaho, and Montana. Arrows show the paths of ancient "megafloods" from Lake Missoula in Montana through the Columbia River into Washington, and from Lake Bonneville into southern Idaho. (Map drawn by B. Lynn Ingram and Wenbo Yang.)

carbonate minerals in lake sediments. These minerals precipitate in the surface waters and eventually settle to the bottom, where they accumulate over time, preserving information about the climate. During dry periods, more water evaporates from the lake surface than is replenished by river inflow, lowering the lake level. Benson reasoned that, since the waters evaporating from the lake are more enriched in oxygen's lighter isotope (oxygen-16), the remaining lake waters should be enriched in oxygen's heavier isotope (oxygen-18). The enriched lake water is then incorporated into the carbonate minerals precipitating in the lake. Therefore, carbonate minerals that precipitate in the lake during dry spells should have a relatively high amount of the heavier oxygen isotope. Conversely, during wetter periods, rain and runoff enter the lake (freshwater that is enriched in the lighter oxygen-16 isotope) is greater than evaporation off the lake's surface; the lake waters and carbonates forming in them are more enriched in oxygen-16 than they are during droughts. Thus, these isotopes in precipitated carbonates provide Benson and other researchers with an archive of past lake chemistry and a record of climate change in the region.

Benson uncovered this long climate history at Pyramid Lake by coring sediments from its bottom. After bringing these cores back to his laboratory in Denver, he separated the carbonate minerals from the lake sediments and dated the cores, finding that the sediments spanned the past 6,500 years. This record revealed that, between 6,500 and 3,800 years ago (the mid-Holocene period), the oxygen isotopes in the lake water were enriched in the heavier isotope, oxygen-18, relative to the lighter isotope, oxygen-16, by an amount indicating that the climate during that time was about 30 percent drier than today, and warmer by 3–5°C (5.5–9°F).

DISAPPEARING LAKES

Benson and colleagues also studied lake sediments from other lakes in the Great Basin, including Owens Lake, to assess the impacts of past climate changes in the Owens Valley east of the Sierra Nevada in California. Like many Great Basin lakes, Owens Lake is a closed basin, receiving inflow only from the Owens River, with no outlet. For many thousands of years, this large lake, twelve miles long and eight miles wide, persisted in the Owens Valley (see figure 18), its size fluctuating in response to changes in the amount of precipitation and evaporation. The lake is now dry—not because of a drier climate but because water managers, starting in 1913, diverted the Owens River south through the Los Angeles Aqueduct to Southern California. By 1924, the lake was completely desiccated, leaving in its place an enormous salt flat composed of clay, silt, and evaporite minerals—the salts that precipitate from the briny waters during evaporation. Beneath this surface, however, lie thousands of years of subsurface sediment layers documenting the lake's response to climate change.

Using the methods he developed for Pyramid Lake, Benson found that Owens Lake had also diminished during the mid-Holocene—between 7,700 and 3,200 years ago. This finding was based partly on oxygen isotope measurements, showing an increase in oxygen-18 in carbonate sediments, and on an assessment of the carbonate mineral types in the lake sediments. Sedimentary carbonates precipitating in lakes generally have two origins: either the carbonate is formed organically during the formation of the shells of organisms that live in the water, or it is formed inorganically, as carbonate precipitates out of the water through chemical reactions. Benson's research at Owens Lake showed that the amount of carbonate in the sediments that formed inorganically (the total inorganic carbon, or TIC), compared with the carbonate that formed organically, was significantly higher during this mid-Holocene period. A higher level of TIC usually indicates that the prevailing climate conditions were drier, resulting in more evaporation off the lake surface, thereby leaving a greater concentration of dissolved salts in the lake waters that precipitate as carbonate minerals. Moreover, the time period between 6,480 and 3,930 years ago was devoid of sediments, and Benson hypothesized that this "hiatus" in sediment deposition resulted from desiccation of the lake during that period, suggesting an extremely prolonged period of drought during the mid-Holocene.

Another large and ancient lake, Tulare, offers its own evidence of a mid-Holocene drought. Located in the southern end of California's Central Valley, Tulare Lake was once the largest freshwater lake west of the Great Lakes until its source of freshwater was diverted for agriculture in the early twentieth century. As recently as 1879, the surface area of the lake was 690 square miles. However, lake sediments and ancient shorelines show that Tulare Lake completely evaporated during the mid-Holocene, again suggesting an extremely dry climate during that period. Researchers have identified imprints of buried mud cracks that developed in the lake bottom as the water evaporated away, and they have studied the fossil remains of land plants that grew on the newly exposed lake bottom, all dating to 5,500 years ago. The lake remained relatively small until 3,600 years ago, when moister climate conditions returned to the region.

Clearly, a large area in the American West was affected by a hotter, drier climate during the mid-Holocene. This climate also altered the vegetation: fossilized pollen preserved in marsh and lake sediments at many locations throughout southern Oregon and extending into the Great Basin and California show shifting plant assemblages. In the mountains, drought-tolerant species, including grasses and shrubs, replaced forests and other densely vegetated areas, and oaks moved to elevations once dominated by pines.

Vegetation change was also seen in coastal regions. In the vast Sacramento–San Joaquin Delta, a freshwater marsh plant community gave way to more salt-tolerant species (saltgrass, pickleweed, and certain types of rushes) as river flows diminished. Along the Southern California coast, including Santa Barbara and the Santa Barbara Channel Islands, drought-adapted marsh plants came to dominate the tidal wetlands.

THE LONG (AND WARM) DROUGHT

This period of prolonged mid-Holocene dryness was coined the "Long Drought" in the late 1930s by Swedish sedimentologist Ernst Antevs, who documented that Lake Summer in south-central Oregon (see location in figure 18) had completely desiccated between 7,500 and 4,000 years ago. This drought in the American West occurred during what is now called the mid-Holocene "climatic optimum"—a period of unusual global warmth. In the early twentieth century, when Antevs was assembling paleoclimatic data, he named this period the "Altithermal." Antevs's research showed that relatively high temperatures, estimated to be 2–9°C (3.5–16°F) warmer than they are today, prevailed during this time in the high latitudes of the Northern Hemisphere. Later studies showed that temperatures at lower latitudes rose less dramatically, to perhaps 1°C (1.8°F) warmer than at present. Antevs's research also suggested that this mid-Holocene period of warm and dry climate was followed by a cool, wet period from about 4,000 to 2,000 years ago, a period we now refer to as the Neoglacial (discussed in the next chapter).

More compelling evidence for increased warmth during the mid-Holocene comes from studies of bristlecone pines in California's White Mountains at their upper limits in the mountains (the tree line, which represents the elevation above which trees cannot survive the cold, inhospitable conditions).

During warmer climate conditions, bristlecone pines are able to survive at higher elevations, and the tree line moves upslope. The radiocarbon ages of ancient tree stumps along the upper elevations of the White Mountains indicate ages of between 5,700 to 4,100 years ago, suggesting a warmer climate. More evidence of warmth was observed in glaciers in the mountain ranges of the western United States that retreated during the mid-Holocene.

COASTAL CONDITIONS

Sediment cores retrieved from along the continental shelf of the Pacific Northwest coast contain the remains of marine organisms that provide clues to past oceanic conditions during the mid-Holocene. These clues include the minuscule fossilized remains of microscopic plants (phytoplankton) living in the sunlit surface waters of the oceans, which convert carbon dioxide and sunlight into energy (in a process known as photosynthesis), providing the primary fuel for the entire marine food web. Some phytoplankton secrete a silica cell wall that cannot be digested by the tiny marine animals (zooplankton) that feed on and then excrete them. As the excrement (or "fecal pellets," filled with these hard, tiny cells) sinks slowly to the bottom, it eventually becomes invaluable data for the oceans' sedimentary archive.

John Barron, a micropaleontologist at the U.S. Geological Survey in Menlo Park, California, has devoted his career to studying the fossil remains of phytoplankton. He is the world's expert on diatoms, a type of phytoplankton that secretes silica cell walls shaped like minuscule hat boxes, footballs, or other fanciful forms—elegant ornamentation that helps them stay afloat in the sunlit surface waters of the ocean by increasing their surface area. Barron has analyzed the different diatom species found in this region and noted that each is adapted to specific ocean conditions, including water temperature, oxygen content, nutrient levels, and water depth.

Barron and his colleagues determined the relative abundances of a dozen or so species found in sediment cores recovered from the continental shelf in the eastern Pacific, from Baja California to Washington state. The oceanographic conditions of this region, just off the coast of western North America, play a major role in influencing climate along the coast, thereby determining its biological resources. Barron's research findings reveal that cold, nutrient-rich waters were upwelling along the coast more often and more extensively between 6,300 and 5,800 years ago. His evidence consists of the increased presence of diatom species (including *Thalassionema nitzschioides*) that live in waters with enhanced upwelling and diatoms (*Neodenticula seminae*) that live in cooler, subarctic waters.

There is an increased concentration of biogenic silica and organic carbon in the sediments (from an increased flux of diatoms remains to the bottom), providing independent evidence for enhanced biological productivity as a result of the increased upwelling of cold, nutrient-rich coastal waters. Moreover, a decrease in the abundance of diatoms (*Pseudoeunotia doliolus*) that live in warm, low-nutrient waters also supports the idea that conditions were cooler and nutrient-rich as the result of increased upwelling. The abundance of nutrient-rich waters would have allowed all marine organisms along the coast to flourish, including those in the intertidal zone that provided food for humans.

Geochemist Timothy Herbert at Brown University, working on the same cores as Barron, was able to make inferences about sea surface temperatures using the fossil remains of the cell walls in a particular species of phytoplankton, *Emiliana huxleyii*. These microscopic plants alter the rigidity of their cell walls in response to changing water temperature. They do this by changing the number of double bonds in their lipid molecules, compounds known as alkenones that behave much like butter, a saturated fat that is a solid at room temperature. (Its structure is "saturated" with hydrogen atoms, so the carbon atoms are held to each other with single bonds.) Unsaturated fat, like oil, is a liquid at room temperature because it has a greater number of double bonds between its carbon atoms. Phytoplankton have evolved the ability to change the number of double bonds in their cell walls to maintain somewhat fluid, or less rigid, cell walls as the water temperature changes. These changes in the ratio of double to single bonds can be measured in the laboratory.

The alkenone results show that coastal surface water temperatures during the mid-Holocene were 1–2°C (1.8–3.5°F) cooler than conditions immediately before or after. The upwelling of cooler, nutrient-rich waters would have led to more prolific phytoplankton blooms, providing more food for organisms higher on the food chain: mollusks, fish, seabirds, and, ultimately, humans.

COASTAL REDWOODS: UPWELLING AND FOG

Some of the most dramatic evidence for increased coastal upwelling during the mid-Holocene comes not from the ocean but from the coastal mountains, where magnificent coast redwood trees grow between southern Oregon and central California. These awe-inspiring trees are the world's tallest living organisms, some towering higher than 360 feet. The coast redwoods require summer fog to survive, and, because of this, they are intimately linked with the oceanographic conditions just offshore in the eastern Pacific Ocean. The fog reduces temperatures during the warm, dry summers and provides moisture for the trees by a process known as "fog drip"—that is, when moisture in the fog condenses on the redwood's fine needles, it drips to the ground where it can soak into the soil and travel to the roots.

Coastal fog is thickest along the Pacific coast during the warm, dry summers. The fog forms when moist air flows inland off the ocean and passes over the cold surface waters of the coast, which causes condensation of the water vapor, resulting in fog. Simultaneously, the warm land surface heats the overlying air, causing it to expand and rise, and, as it rises, that warm air is replaced by the cool marine air and fog flowing onshore. In regions of the coast with small mountain ranges, this onshore movement of fog can appear like a cloud waterfall pouring over the coastal mountains, bringing precious moisture to coastal ecosystems during the dry summer months.

The Pacific coastal redwoods form a geographic boundary, or ecotone, between the more arid oak woodland and open scrub of Southern California and the wetter western hemlock, spruce, and alder lowland coastal forests of Oregon and Washington. An analysis of pollen from cores taken along the California and Oregon coasts suggests that coastal upwelling and the development of summer fog were coincident with a great expansion of these majestic redwood forests that began about 5,150 years ago.

HUMAN MIGRATIONS

Against this backdrop of a warming and drying climate in the West during the mid-Holocene, it is clear that the early humans living in the region faced tough times. Archaeologists have found evidence that humans were likely suffering from the severe droughts, especially in southeastern California and the Great Basin. Prolonged and severe droughts appear to have forced large-scale migrations out of the driest inland regions as people went in search of food and water. Douglas Kennett, an archaeologist at Penn State University who has studied these migrations, reasons that the moderate climate and abundant marine resources along the coast would have attracted native human populations from the drier interior deserts. Kennett has collaborated with his father, paleoceanographer James Kennett (see figure 19), to reconstruct coastal sea surface temperatures in the Santa Barbara Basin during the Holocene. Climate conditions along the coast were not as severe as inland, and the high biological productivity along the coast would have provided marine resources not available in the desert interior. The wetlands of California's Central Valley would also have provided more resources and favorable conditions than the interior desert regions. At a time when the Great Basin and southeastern California were severely dry, the coast would have offered increased food and water resources.

According to Douglas Kennett, archaeological evidence supporting a coastal migration during the mid-Holocene comes from a study of the distribution of language groups in western North America. In particular, the Uto-Aztecan language family is distributed between coastal regions of present-day Southern California, the Southern California desert, and much of the Great Basin. This distribution suggests a movement of people from the interior to the coast.

Other archaeological evidence suggests that populations from the dry desert interior increased their interactions and trade with peoples along the California coast and Central Valley during the mid-Holocene drought. This evidence is contained in shell mound deposits excavated along the central and southern coasts, including the Santa Barbara Channel Islands. Exchanges among the tribes may have buffered the impacts of the drought, lowering the risks of reduced resources that would have accompanied extended dry conditions. Evidence for increased trade comes from beads crafted from the marine shell *Olivella biplicata,* or "purple olive snail." During the mid-Holocene, rectangular beads were cut from the body whirl of these snail shells and used for trade and exchange. Specialized tools (micro-drills and anvils) for fashioning these beads, as well as the debris from their manufacture, were recovered on the Santa Barbara Channel Islands. The beads have been found in sites along the Southern California coast and in the northern and western Great Basin. Based on the ages and contexts of these beads, their distribution suggests that new trade routes were established between 5,900 and 4,700 years ago—the driest period of the mid-Holocene.

EXPANDING COASTAL ENVIRONMENTS

The coastal environment was just beginning to stabilize starting about 6,000 years ago, in the midst of this mid-Holocene drought. The period that

FIGURE 19. Archaeologist Douglas Kennett (on left) with his father, paleoceanographer and paleoclimatologist James Kennett, resting after fieldwork on San Miguel Island in the Santa Barbara Channel off the Southern California coast. (Photo courtesy of Douglas Kennett, Penn State University.)

followed the end of the Last Glacial Maximum, or 20,000 years ago, was marked by the melting of the massive glaciers that had shrouded much of the Northern Hemisphere. The retreat of the glaciers started slowly, but, as more land was exposed, the bare ground and open waters absorbed more solar radiation than did the glacial ice sheets, warming the earth's surface. This led to an increase in the rate of melting. Large volumes of the water from the melting ice sheets flowed into the world's oceans, raising global sea level at a rapid pace and transforming coastal regions worldwide.

Along the California coast, San Francisco Bay began its transition from a river valley 18,000 years ago to a large estuary system 8,300 years ago as rising sea levels began to enter the bay through the Golden Gate. This rapid sea level rise continued until the mid-Holocene, about 6,000 years ago, when the rate slowed to only about an eighth of an inch per year and the shoreline stabilized. This allowed coastal habitats along the Pacific Ocean, including tidal marshes, the rocky intertidal zone, beaches, bays, and estuaries (including San Francisco Bay) to develop and expand. These newly created coastal habitats would have provided new sources of abundant marine resources. Douglas Kennett proposed that some of the Uto-Aztecan people who had migrated from the dry interior deserts to Southern California then migrated northward, through the Central Valley of California and into the wetland environments of the San Francisco Bay region. The oldest shell mound site along the shores of San Francisco Bay—the West Berkeley shell mound—has a basal age that dates to the mid-Holocene, or 5,200 years ago.

Along the Pacific Northwest coast, the basal dates of a number of shell mounds are mid-Holocene in age, suggesting that coastal populations expanded there as well. Food resources expanded in that region, as they did along the southerly California coast, when sea levels stabilized. The newly formed intertidal zone would have brought with it the expansion of resources, such as shellfish, which allowed human populations to grow and construct the numerous shell mounds. Conditions also improved for salmon migration, spawning, and development of juvenile salmon. Archaeologist Knut Fladmark at Simon Fraser University proposed that an increase in the abundance of sockeye and coho salmon, considered the single most important Native American food resource in that region, led to an increase in coastal settlements.

The climatic backdrop of these changes based on evidence from the Pacific Basin suggests that the mid-Holocene was a period of increased La Niña conditions, when the trade winds blew with more intensity. It is possible that the La Niña conditions over the Pacific were at least partly responsible for the extended period of drought over the American Southwest.

Following the mid-Holocene drought, climate in the American West turned cool and wet again during a period known as the Neoglacial. The archaeological record contains evidence for increased human sedentism, cultural complexity, and more advanced technology beginning about 4,900 years ago, becoming even more widespread after 3,800 years ago. As we will explore further in the next chapter, the Neoglacial brought with it increased climatic variability.

EIGHT

Ice Returns

THE NEOGLACIATION

> We are living in an epoch of renewed but moderate glaciation—a
> "little ice age" that already has lasted about 4,000 years.
>
> FRANÇOIS MATTHES, *Report of*
> *Committee on Glaciers*

THE WARM, DRY CONDITIONS of the mid-Holocene gradually gave way to a
cooler, wetter period known as the "Neoglaciation." It had no clearly demarcated beginning, settling unevenly over the Northern Hemisphere with
considerable local variation. In Europe, notably in southern Norway and in
the Swiss and Austrian Alps, the onset of this cooler period occurred 5,000
to 4,000 years ago, based on the dating of glacial moraines and of sediments
that were eroded by glaciers and transported by glacial meltwaters. Glacial
expansion across Iceland, beginning 5,000 years ago, and the retreat of the
Eurasian tree line between 4,000 and 3,000 years ago also signaled cooler
temperatures.

In the mountain ranges of the American West, many cooler summers
passed without melting the previous winter's snow accumulation. At a
pace of tens of feet per year, glaciers in the high mountain ranges of the West,
from Alaska to western Canada and into the Cascades, Sierra Nevada, and
Rockies, grew and flowed downslope, and the tree line migrated to lower
elevations.

This period of Northern Hemisphere cooling also brought wetter conditions to the American West. Lakes formed in unlikely places, including some
of the driest places in the West today, such as the Mojave Desert of southeastern California. When researchers cored into the dry Mojave lakebeds,
or playas, they were surprised to find evidence that large ephemeral lakes,
including Silver Lake today, had formed there 3,620 years ago. In nearby
Searles Basin, playa lakes formed about 3,590 years ago.

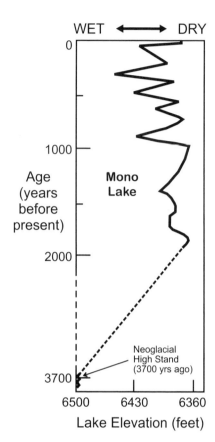

WET ⟵⟶ DRY

Mono Lake

Age (years before present)

Neoglacial High Stand (3700 yrs ago)

Lake Elevation (feet)

FIGURE 20. Ancient levels of Mono Lake over the past 3,800 years, reconstructed by using the deltaic sediments of inflowing streams. Note the high stand 3,700 years ago. (Based on data from Stine 1990.)

In eastern California, more evidence for a wet Neoglacial climate was found in Mono Lake in the Owens Valley, to the east of Yosemite National Park. Mono is a special lake, not just for its unique ecosystems and fascinating tufa towers but also for its great age: the lake has existed for over a million years, making it one of the oldest lakes in North America. The sediments accumulated beneath Mono Lake therefore contain a rich archive of past natural history. The lake is a "closed basin," meaning water flows into it but has no outlet. During dry periods, when the rate of evaporation from the lake exceeds inflows from direct precipitation and streams, the water level subsides. During wetter times, when more water flows into the lake than evaporates, the lake expands. In this way, Mono Lake acts as a prehistoric rain gauge. Several lines of evidence indicate that the lake swelled to its highest known level (or high stand) some 3,730 years ago (see figure 20).

Owens Lake in the Owens Valley south of Mono Lake, having desiccated during the prolonged mid-Holocene drought, also reached a high stand during the Neoglacial. In the Sierra Nevada range, Lake Tahoe, which had been unusually low for hundreds of years during the mid-Holocene, was restored to its early Holocene high level by about 3,000 years ago. The lake again reached the elevation of its sill, causing lake waters to overflow into what is now known as the Truckee River, its only outlet. These waters flowed eastward into Nevada, eventually reaching Pyramid Lake, which also reached a high stand about 3,000 years ago.

Fossilized pollen and plant remains in lakes and meadows in the Sierra Nevada provide further evidence for cooler and wetter climatic conditions. For instance, pollen from Lake Moran in the Sierra Nevada shows a shift in vegetation to cool-adapted fir and pine trees in the region surrounding the lake, starting around 4,000 years ago. An analysis of charcoal levels in Sierra lake sediments shows that these levels declined during the Neoglacial, indicating reduced fire frequency.

CALIFORNIA'S CENTRAL VALLEY

More snow and rain in the Sierra Nevada would have caused more serious flooding along the rivers draining the range, including the Sacramento and San Joaquin, as floodwaters would overtop the banks of the rivers, spread out over the floodplains of the Central Valley, and fill the natural wetlands that once covered the region. Evidence that major flooding occurred in the Central Valley during the Neoglacial period is found in Tulare Lake, located in the southern San Joaquin Valley. Prior to its recent human-caused desiccation, Tulare was the largest freshwater lake in the American West, receiving inflow from the Kings, Kaweah, Tule, and Kern rivers that drain the central and southern Sierra Nevada. During wet periods, these rivers brought even more water to the lake, expanding Tulare and its surrounding wetlands.

Paleoclimate researchers analyzed pollen extracted from Tulare Lake sediments to assess the extent of marshes in the Tulare Lake Basin by determining the proportion of aquatic plants relative to saltbush, a plant that grew around the lake's shore. These sediment studies, along with shoreline analyses, show that Tulare Lake experienced a high stand between 4,000 and 2,700 years ago. The increased high water and flooding during this generally wetter period would have created conditions much like those seen in 1861–62, when the Central Valley was transformed into an inland sea. These earlier floods would have destroyed any human settlements in the Central Valley. For this reason, it is likely that the native populations used the Central Valley only seasonally and not as a location of permanent settlement.

MOUND BUILDERS OF THE SAN FRANCISCO BAY

Downstream of the Sierra Nevada and the Central Valley, rivers draining these regions flow into San Francisco Bay before reaching the Pacific Ocean. Clues to the climate of the Bay Area during the Neoglacial are found in sediment accumulation beneath the bay waters as well as in archaeological investigations of the shell mound villages along the coast.

The remains of several hundred mysterious mounds that once served as dwellings for a population of hunter-gatherers along the shores of San Francisco Bay have proven to be valuable—if unlikely—sources of information about Neoglacial climate in the West. Archaeologists discovered these mounds in the early twentieth century, along with their most unusual feature: they were composed almost entirely of shellfish remains.

Unlike mounds found elsewhere in North America, which were primarily specialized burial sites, the Bay Area mounds were large enough to suggest that they functioned as residences. Because they were created through the daily accumulation of household debris, the mounds contain a wealth of materials for reconstructing the environment and the climate, including the remains of fish and shellfish from the bay, the bones of terrestrial animals, and a wide array of foraged remains.

But before the significance of the mounds could be understood, archaeologists in the early twentieth century were forced into a race with urban development. One of the leading archaeologists of the time, Nels Nelson, worked heroically to map over 425 of the mounds, but he noted as early as 1909 that most of the larger mounds had been disturbed. Archaeological fieldwork in the Bay Area became largely a salvage operation, with scientists frantically digging ahead of the bulldozers that were poised to level the sites for new roads, parking lots, and businesses. During these rushed excavations, there was barely time to catalog artifacts before whisking them away for storage in museums. Except for those fragments (and a few shell mounds protected intact within state parks), the archaeological record of the earliest human inhabitants of the San Francisco Bay region has been largely erased.

Despite this enormous loss, however, the achievements of Nelson and his successors have helped us piece together a more complete picture of human life during the Holocene, including the Neoglacial. As the droughts of the mid-Holocene gave way to the wetter Neoglacial, the pace of mound-building increased, as shown by radiocarbon dating of successive shell and soil layers of the West Berkeley shell mound. The so-called "Middle Period," from about 4,000 to 1,450 years ago, was thought to have been a "golden age" of shell mound activity, with the number of mounds around the Bay Area reaching a peak in terms of population numbers and year-round occupation.

If we visualize San Francisco Bay during that time, the curious mounds scattered around its margins were oblong—up to 500 feet long and 30 feet high—and would have been dominant features of the shoreline. We might see smoke curling up from village fires as the people rested from their foraging and prepared their meals. They had for many centuries made the marshes their home for part of each year after sea levels had stabilized 6,000 years ago, allowing the expansion of mudflats and tidal marshes. These, and the rocky intertidal zone and associated ecosystems, supported an abundance of coastal resources. The people caught fish and harvested mussels, clams, and oysters from the shores of the bay. They used rushes that grew close to the channels and in the fresher parts of the marsh for building their homes and making baskets to carry implements and food. Their drinking water was obtained from local creeks that drained the nearby hills, and they foraged riparian forests surrounding the creeks for berries, roots, and herbs. They hunted deer, rabbits, birds, and other game animals.

There were challenges to permanent habitation, however. The Neoglacial period was marked by more frequent rains and increased river flows that raised the shoreline, and storms threatened coastal settlements that were unprepared for them. The shell mound villages represented a creative response to this challenge: residents began "piling up" rather than "carving out" a living. As kitchen waste from each household, including the remains of cooking fires, was collected and deposited outside the simple huts, it became part of a continually rising ground surface. The ground and the mounds rose steadily, fast enough for the villages to keep up with the rising tides that each year reached a little higher onto the marsh. The recycling habit of the mound builders was not just good housekeeping but also an essential community service and survival tool. These mounds therefore contain a stratified archive of the lifeways, environments, and climates of the region over the past several thousand years, and they suggest that the climate was indeed wetter, with higher rates of river inflow, during the Neoglacial.

Sediments that have accumulated beneath San Francisco Bay also contain evidence for prolonged wetness and occasional extreme events during that period. Variations in freshwater inflow alter the bay's salinity over periods of decades to centuries and affect the organisms living in the estuary—from single-celled plankton and foraminifera to mollusks (oysters, clams, and mussels). The assemblages of species living in the bay adjust in response to changes in river inflow: during higher river inflow, species that prefer lower salinity predominate, whereas during periods of decreased inflow, such as would occur during drought, species that prefer higher salinity conditions predominate. When these organisms die, their hard shell remains settle to the floor of the bay and are entombed in the sediments brought in by creeks and rivers. These accumulated sediments provide a source of environmental information for paleoclimatologists.

In addition to the identification and analyses of bay fossils, the geochemistry of these shells reflects the environmental conditions of the bay water, particularly the salinity and temperature. Such factors in turn reflect changes in the precipitation regime over the bay's vast watershed that covers about half of the state of California today (see figure 21). This means that the chemistry of the bay water can inform us about climate over a very large region. As shellfish grow and secrete their calcium carbonate shell, they incorporate elements from the water surrounding them, including oxygen. Not all water is the same. Tiny variations in the mass of oxygen atoms (actually, isotopes) have been exploited by geochemists to provide information about past environments. River water entering San Francisco Bay through the delta is lighter than the Pacific Ocean water that enters the bay through the Golden Gate in that it has a greater proportion of oxygen-16. The differences in the amounts of fresh- and saltwater in the bay determine its salinity. The amount of river water flowing into the bay varies with the seasons, with more river water flowing during the wetter winter months. During years and decades marked by higher rainfall, the average salinity in the bay is lower than it is during years and decades of drought.

During the late 1980s and early 1990s, a research group including this book's authors, in collaboration with other scientists researching San Francisco Bay at that time, collected and analyzed samples of bay waters during different seasons over several years, measuring the salinity and oxygen isotopes of these waters. The samples provided evidence that both the seasonal and the annual average salinity fluctuations were reflected in the proportions of oxygen-16 and oxygen-18 in the bay water. We then collected modern

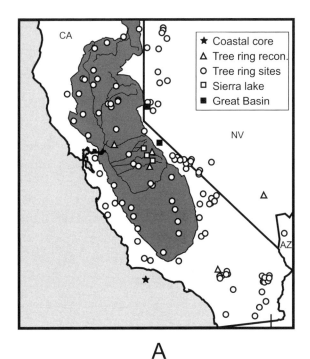

A

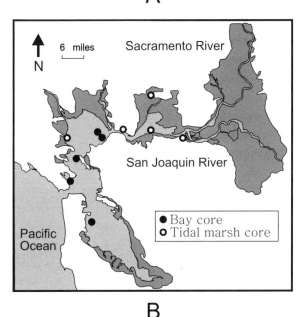

B

FIGURE 21. Maps of California and the Great Basin showing many of the locations of paleoclimate records discussed in chapters 8–10. In map A the watershed of San Francisco Bay is shaded. Map B of San Francisco Bay shows locations of cores taken from beneath the estuary and surrounding marshes for paleoclimate studies. (Maps by Frances Malamud-Roam.)

mussel shells growing in parts of the bay of known salinity and were able to show that the oxygen isotopes of these shells reflected the salinity of the bay.

These calibration studies were done to support the analyses of sediment cores taken by our research group from the central bay (see figure 21). The cores provide a record that extends back in time more than 5,000 years. Once collected from the bay, the cores were brought to the laboratory, where fossilized shells were separated in sequence from the sediments every two centimeters (three-quarters of an inch). The shells were processed, including cleaning to avoid contamination from the surrounding sediment and powdering to homogenize the samples. A tiny amount of this powder was then analyzed using a mass spectrometer to measure the proportions of oxygen-18 relative to oxygen-16.

The results of these measurements showed that the San Francisco Bay estuary was much fresher during the Neoglacial period. The decrease in salinity suggests that the average amount of river inflow was twice as high as the modern average inflow. The flows peaked between about 4,300 and 3,800 years ago, and average river flows remained relatively high for the following 2,000 years.

We were able to apply newly developed geochemical methods for assessing river inflows and salinity in San Francisco Bay to the shell mound collections. Our collaborators, archaeologists Kent Lightfoot at the University of California, Berkeley, and Ed Luby of San Francisco State University, have spent much of their careers researching the purpose of these mounds, including when and why they were occupied. They provided access to shell mound materials housed in the Hearst Museum of Anthropology on the UC Berkeley campus. The animal remains in some Bay Area mounds had been investigated to assess the season of occupation by analyzing the proportion of migratory species that were hunted by the native populations. For instance, more than half the bird remains in a mound from southern Marin County were migratory species that were present only during the fall and winter. Two large mounds—the West Berkeley and Emeryville shell mounds on the eastern shore—contained abundant chinook salmon remains, suggesting an occupation during the spring and fall migratory periods.

Our geochemical methods would allow us to assess the season of occupation in other mounds that did not contain these types of faunal remains and thus could potentially be applied to a larger number of mounds throughout the Bay Area. We decided to test our methods on two mounds located along the eastern shore of the central part of San Francisco Bay: Ellis Landing, a large shell mound close to present-day Richmond, and a small mound located on a tiny nearby island (Brooks Island).

A graduate student working in our group, Peter Schweikhardt (now a professor at the College of Alameda), developed a novel way of using fossilized shell fragments to determine the season of harvest. The results showed that shells from the Ellis Landing mound were harvested mainly during the spring and fall seasons. Some appeared to have been harvested during the summer, but none were harvested during the winter. This suggests that these mounds were used seasonally during the spring and fall but were largely abandoned during the summer and winter. The smaller mound from Brooks Island also showed seasonal occupation during the fall to early winter, though this mound appeared to have been used more during the summer as well, starting in July. The native populations apparently followed the water and other resources in the region throughout the year as the weather shifted between the cold, wet winter season and the warm, dry summer season. As we shall see in later chapters, this mobility served them well when the climate again shifted to drier, warmer conditions.

A MORE VARIABLE CLIMATE

The oxygen isotopic measurements from the San Francisco Bay sediment cores also show that the abundance of freshwater was part of an overall pattern of climate fluctuations between wetter and drier conditions occurring about every 1,500 years. This climate cycle has appeared elsewhere in the West. For instance, researchers studying Tulare Lake have determined that the lake experienced seven or eight high stands over the past 11,500 years, spaced approximately 1,500 years apart, and coastal sea surface temperatures reconstructed from oxygen isotopic measurements of foraminifera from Santa Barbara Basin sediment cores show a 1,500-year cycle, as we discussed near the end of chapter 6.

In the bay sediment cores, we also detected fluctuations that were smaller and more frequent, with 200-, 90-, and 55-year periods. We will return to these cycles in later chapters, since they appeared in a number of paleoclimate records in the West, as well as globally, and may be related to sunspot cycles and associated changes in solar radiation.

Relative to the early and mid-Holocene, the climate of the late Holocene period—beginning about 3,500 years ago—exhibited larger and more rapid changes. In other words, the climate became more variable. From the coastal ocean to the inland Great Basin, the evidence all points to a less stable climate. Along the coast, for instance, in the Santa Barbara Basin, sea surface temperatures, upwelling, and the thicknesses of the sedimentary layers show greater variability. Pollen separated from marine sediment cores taken off the Central California coast suggests that vegetation growing in those regions began to show more rapid fluctuation from wetter to drier species starting about 3,400 years ago. Inland at Pyramid Lake, in western Nevada, studies of pollen and plankton (diatoms) from lake sediment cores also exhibit an increase in climate variability after 3,500 years ago. This increased variability may be due to an increase in the frequency of El Niño events that appears to have begun at that time. A more detailed discussion about changes in the ENSO and other climate cycles will be given in chapter 11.

The wetter and cooler climate of the Neoglacial gave way to a period of drying conditions after about 1,800 years ago. Was this change favorable or unfavorable for our intrepid mound builders? As we shall see in the next chapter, there is evidence that these mound dwellings, as well as other Native American settlements in California and elsewhere in the West, were abandoned during this warmer and drier period, perhaps in response to the change in climate.

NINE

The Great "Medieval Drought"

When we move beyond Europe and the North Atlantic into
drier environments and lands with unpredictable rainfall, we
enter a Medieval world where drought cycles and even a few
inches of rain could make all the difference between life and
death. While Europe basked in summer warmth and the Norse
sailed far west, much of humanity suffered through heat and
prolonged droughts. Archaeology and climatology tell us that
drought was the silent killer of the Medieval Warm Period, a
harsh reality that challenged human ingenuity to the limit.

BRIAN FAGAN, *The Great Warming*

THE COOL, MOIST CONDITIONS of the Neoglacial period ended in the
late Holocene, with the climate becoming drier and warmer beginning about
1,850 years ago and continuing for more than 1,000 years. Although the general trend in climate in the modern-day western United States was toward
dryness, conditions also became more variable: prolonged droughts were
interspersed with sudden excursions of extreme wetness. The paleoclimate
record of the past 1,800 years contains many examples of such climate swings,
and, though both ends of the climate spectrum are of concern, it is the pros-
pect of extended drought that most worries water planners in the West today.

A period that included episodes of extended and severe dryness began in
about AD 900 and lasted until about AD 1400. Sometimes referred to as the
"Medieval drought" in the western United States, this period was marked (in
some regions of the West) by two prolonged "megadroughts" separated by a
wet interval around AD 1100–1200. These droughts are considered by many
paleoclimate investigators as possible analogs of the protracted droughts the
West could face in the future and, as such, they provide an important impe-
tus for water sustainability planning.

We begin this chapter by looking at the demise of the Ancestral Pueblo
civilization in the Southwest, which occurred near the end of the Medieval
drought period. This example shows how the megadroughts affected early

human societies of the region. Our attention then turns to the physical evidence for the megadroughts and the impacts they had on the ecosystems and coastal populations of the West.

THE ANCESTRAL PUEBLO COLLAPSE

An advanced agricultural society once flourished in the American Southwest, over a thousand years before air conditioning and massive public water works made this region hospitable to our modern society. How could such a hot, dry region have become a vibrant center for ancient societies without the technological advantages we enjoy today? More curiously, why did this society apparently collapse at its peak of sophistication and population? Since first discovering the ruins of the Ancestral Pueblo—the collective name given to the early people of this region formerly known as the Anasazi— scientists have struggled to answer both questions.

Much of what archaeologists know about the Ancestral Pueblo comes from pueblo and cliff dwelling sites from the Four Corners region, including Chaco Canyon in northwestern New Mexico, Mesa Verde in southwestern Colorado, and Canyon de Chelly in northeastern Arizona. Archaeologists have used the deteriorating remains of adobe pueblo walls, artifacts, pottery shards, preserved food, soils, clothing, animals, and the skeletal remains of the people themselves to reconstruct this culture and better understand its collapse.

Chaco Canyon in New Mexico was the site of one of the most extensive Ancestral Pueblo settlements. By the ninth century, a warmer and wetter climate allowed populations in this region of the Colorado plateau to expand. At its peak, during the eleventh and early twelfth centuries, Chaco Canyon was elegant, with great pueblos the size of modern apartment buildings housing hundreds of residents in large, high-ceilinged rooms. Enclosed within these great houses were plazas and special circular sunken rooms, called kivas, where people slept and worked. The Ancestral Pueblo settlements were cities in many respects, supported by agriculture, which had begun as early as the second century and allowed people to settle in one place year-round. These early farmers planted corn and a variety of vegetables, and, over the decades and centuries, agriculture expanded onto the higher mesas to feed the growing population. Most of this farming depended on annual rains, supplemented by water from nearby streams and springs. Over time, however, the climate became increasingly arid and unpredictable; the Ancestral Pueblo farmers were forced to develop ways to divert water from rivers and even devise ways of storing it for future use. The residents of Chaco Canyon built an extensive system of diversion dams and canals, directing rainwater from the mesa tops to fields on the canyon floor, allowing them to expand the area of arable land.

With the help of this early water development, Chaco Canyon grew into a complex center of Ancestral Pueblo culture during the tenth and eleventh centuries. Water diversion features found in Chaco Canyon include terraced cliffs, ditches, overflow ponds, and masonry irrigation canals and dams. The Ancestral Pueblo were primarily agricultural, depending on their primary crops of corn, beans, and squash, which were supplemented with occasional hunting and the gathering of wild plants.

For centuries, the Ancestral Pueblo society thrived, closely connected to the natural climatic rhythms of the region. The summer monsoons provided sufficient water during the growing season, and, in the occasional years when the monsoons failed, residents relied on food reserves kept in large storage rooms in the great houses. The population in the Four Corners region swelled throughout the eleventh and twelfth centuries—until things began to go wrong.

Farther to the west in southern central Arizona, the Hohokam people settled near the confluence of Arizona's only three rivers—Gila, Verde, and Salt. This culture initially practiced floodwater farming on the Gila and Salt rivers. Each year, they planted crops in the wet floodplain soils after the spring floods overtopped the riverbanks and spread water and nutrient-rich silts and clays in a manner analogous to ancient farmers thousands of miles away along the River Nile.

Within fifty years or so, the Hohokam had developed a new technology, canal irrigation, which began with the construction of small canals close to the rivers for transporting water to the fields. The Hohokam civilization thrived in central Arizona for a thousand years, building an extensive network of integrated canal systems that were capable of transporting large volumes of water long distances. The largest canals were twenty-four feet wide, nineteen feet deep, and twenty miles long—capable of irrigating over 10,000 acres of land. The engineering of these complex canal networks is considered to be the Hohokam's greatest achievement, and the irrigation systems of the Hohokam are considered the best in prehistoric North America. Their success allowed them to expand across a large region of Arizona, from modern-day Tucson to Flagstaff, a distance of 260 miles. At the peak, an estimated 40,000 Hohokam lived in Arizona, but they suddenly vanished in the mid-fifteenth century.

FIGURE 22. Montezuma's Castle, a cliff dwelling located just north of Camp Verde in central Arizona. (Photo by B. Lynn Ingram.)

Another culture, the ancient Sinagua (which translates to "without water"), flourished farther to the north in Arizona, in the region between Phoenix and Flagstaff today. The Sinagua initially controlled water for their crops using check dams—small walls built of stone across ephemeral streams—and irrigation ditches. Over time, sediment and water accumulated behind these dams, providing fertile soil for planting crops. The Sinagua built storage reservoirs for domestic uses of water, such as washing, cooking, pottery making, and drinking. They also took advantage of natural water reservoirs, thriving in an arid environment for a thousand years. They built cliff dwellings near Camp Verde, Arizona (such as Montezuma's Castle), which suggests that, as resources became scarce, they were forced to build fortified dwellings with hidden food storage areas to protect their meager supplies (see figure 22). The Sinagua also disappeared very suddenly, at about the same time as the Hohokam.

Looking closely at the demise of these ancient civilizations, we find many similarities. In particular, all of these societies were flourishing prior to a rather abrupt collapse. The archaeological record of the last decades of the Ancestral Pueblo in Chaco Canyon abounds with evidence of suffering: skeletal remains show signs of malnutrition, starvation, and disease; life spans declined; and infant mortality rates soared. Chilling stories of violence, possibly warfare, were told by mass graves containing bones penetrated with arrowheads and teeth marks and skulls bearing the scars of scalping.

Piles of belongings were found, apparently left behind as the people abandoned their settlements and fled, some to live in fortified hideouts carved in the cliff faces, protecting their hoarded food from enemies. The population of the region is known to have plummeted by 85 percent in a short span of time, but the causes of the collapse are still debated. Many make the case that climate was clearly a factor. In this chapter, we explore some of the climate evidence for drought, beginning with marshes along the California coast.

DROUGHT EVIDENCE

San Francisco Bay

Vast tidal marshlands have grown up around the edges of San Francisco Bay for roughly 6,000 years, after sea levels began to stabilize. The slower rate of sea-level rise allowed marshes to establish themselves and keep pace with the encroaching water. As the marshes continued to grow, their annual remains accumulated, decomposing slowly in the anaerobic conditions below the surface and forming peat. These marsh peats are important climate archives for scientists, safeguarding information about climatic and environmental conditions over the millennia.

The San Francisco Bay is like a "funnel" for water draining Northern and Central California out to the Pacific Ocean. The bay and its watershed can be seen in dramatic clarity in satellite images, which show a web of river tributaries from the Coast Ranges and the Sierra Nevada to the Central Valley toward the bay. Fifty miles inland from the Golden Gate, these rivers converge to form the Sacramento–San Joaquin Delta, where the waters mingle and flow through the bay and out Golden Gate to the Pacific. Within San Francisco Bay, fresh river water meets and mixes with the incoming ocean water, producing a range of salinity: fresh in the delta, saline in the central part of the bay near the Golden Gate, and brackish (intermediate) in between.

Brackish waters provide unique habitats for plants growing in adjacent tidal marshlands, creating more diverse plant communities than either the salt marshes near the Pacific Ocean or the freshwater marshes of the delta. The plants and organisms living in these tidal marshes must be adaptable: the daily tides bring in not only more water but more *salty* water. Furthermore, the annual march of the seasons brings changes in overall salinity as well, particularly in this region marked by wet winters and dry summers.

These annual cycles are superimposed on the year-to-year variability in salinity, reflecting variations in climate patterns. A drought in the watershed, if prolonged and severe, can cause higher salinity downstream in the estuary as the inflow of freshwater drops. In response, salt-tolerant species in the marshes expand further inland toward the delta, and the freshwater species retreat. Conversely, unusually wet winters generate fresher conditions in the estuary, leading to an expansion of freshwater-adapted species.

On the basis of this relationship between rainfall and wetland species in the estuary marshes, Roger Byrne, a Geography professor at the University of California, Berkeley, reasoned that a history of marsh plant assemblages could be used as a proxy record of past climate conditions, particularly for droughts and wetter periods experienced over the whole watershed. Such a history of vegetation could be produced from the evidence buried in the marsh sediments, which have accumulated abundant organic plant remains over thousands of years. Byrne invited the research group of this book's authors to collaborate on this ambitious project, and together we cored marshes along the edges of San Francisco Bay across a range of salinities between the delta and the Golden Gate, with the goal of assessing past climate regimes over California as recorded by the marsh ecosystems (see figure 21 in the preceding chapter).

A number of proxy climate indicators were used in the project. For instance, pollen from the marsh sediment cores was used to assess past vegetation change. This pollen was extracted from the sediments, in addition to seeds and other vascular plant remains that had been deposited on the marsh surface over the millennia. Snapshots of the ancient plant assemblages were created from each layer, providing clues about the climate.

Suisun Marsh, located midway between the delta and the Golden Gate, provides a particularly sensitive record of climate change. By AD 1850, the wetlands surrounding Suisun Bay represented the most extensive area of brackish marsh on the west coast of North America. These marshes have experienced rapid losses since that time, as over 90 percent of Suisun Marsh has been filled for urban development and agriculture. Our study site in the northern part of Suisun Bay, called Rush Ranch, was a small, relatively pristine relict of the Suisun tidal marsh. By assessing the pollen remains in the sediment cores, we could determine the percentage of marsh plants adapted to higher salinity—such as saltgrass (*Distichlis spicata*) and cordgrass (*Spartina foliosa*), the so-called Chenopodiaceae—relative to those adapted to fresher conditions, like tules, cattails, and rushes, to assess the salinity conditions in past periods when the plants were growing. We knew that the marsh vegetation responds quickly to the changing salinity of freshwater inflow.

Our research project also involved the measurement of carbon isotopes from partially decomposed plant remains in the marsh cores as another way of deciphering changes in marsh vegetation. Nearly all carbon atoms have an atomic mass of 12 (carbon-12), but a small number of them have a mass of 13 (carbon-13). During times of drought, Suisun Bay would have been a saltier place, because less freshwater would have entered from the river. With higher salinity, the two salt-hardy marsh species, saltgrass and cordgrass, increased in dominance. Both of these species have higher proportions of carbon-13 relative to carbon-12, so their remains in the sediments could be detected using a mass spectrometer to measure the relative proportions of carbon isotopes.

Another important element of the research project included a third proxy tool: diatoms. Diatoms are microscopic phytoplankton that live in all aquatic environments, including marshes, and secrete a minuscule silica frustrule. Each diatom species produces a uniquely shaped frustrule, and these accumulate in the marsh sediments over time. Some species of diatoms prefer a more freshwater environment, whereas others prefer a more saline environment. By looking at the relative proportions of these diatoms in the sediment cores, our collaborator, paleontologist Scott Starratt at the U.S. Geological Survey, was able to provide an independent measure of water salinity in the marsh over time.

We dated the Suisun marsh core sediments using radiocarbon (carbon-14) measurements on the seeds of marsh plants separated from various levels of the core. This allowed us to establish a chronology delineating the vegetation changes detected through the pollen, diatom, and isotope analyses.

The results were compelling. They revealed that the salinity of the marsh increased (meaning the average freshwater inflow to San Francisco Bay was reduced) by close to 40 percent above today's levels for a thousand years, between 1,750 and 750 years ago. The peak of this salty, low-inflow interval occurred approximately 1,200 to 900 years ago, during the early part of the Medieval drought. Only very low precipitation and runoff over San Francisco Bay's vast watershed, covering almost half the area of California, could explain these results (see figure 23).

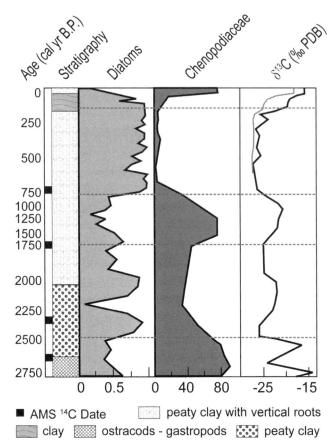

AMS ¹⁴C Date   peaty clay with vertical roots
clay   ostracods - gastropods   peaty clay

FIGURE 23. Stratigraphy of the sediment core from Rush Ranch, Suisun Marsh, in the northern San Francisco Bay. From left to right are the calibrated radiocarbon ages, the sediment types in the core, the diatom index, the percentage of Chenopodiaceae pollen, and carbon isotope values. The diatom index refers to the percentage of diatoms that live in higher salinity waters versus fresher waters. Values ranging between 0.00 and 0.30 are dominated by marine taxa, 0.31 to 0.70 are dominated by a mixture of taxa, and 0.71 to 1.00 are dominated by freshwater taxa. The Chenopodiaceae pollen (from marsh plants that grow in higher salinity waters, such as *Salicornia* and *Atriplex*) represents the percentage of that pollen relative to the total marsh pollen. The d13C value δ¹³ (plotted on the right) of fresher marsh plants is closer to -25 per mil, with those growing in higher salinity waters being closer to -15 per mil. The data show a higher salinity interval from 1750 to 750 years ago, encompassing the Medieval Climate Anomaly. (Graph modified using data from Byrne et al. 2001.)

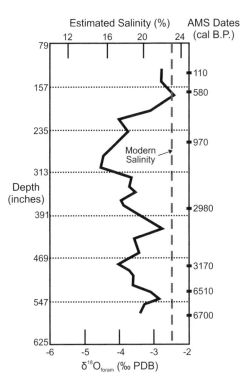

FIGURE 24. Oxygen isotope record from foraminifers from San Francisco Bay sediments, showing a shift to higher bay salinity (in parts per thousand) starting about 1,800 years ago and peaking 600 years ago. (Data from Schweikhardt, Sloan, and Ingram, 2008.)

Further evidence from San Francisco Bay sediment cores has been found for the major shift to a drier climate in California. The oxygen isotopic ratios (oxygen-18/oxygen-16) of calcium carbonate shells of microscopic, single-celled organisms (foraminifera) vary with salinity in the bay, as described in chapters 5 and 8. A research group that included Peter Schweikhardt and one of this book's authors, B. Lynn Ingram, retrieved sediment cores from the central part of San Francisco Bay near Point Richmond. Along with micro-paleontologist Doris Sloan, we chose one specific species of foraminifera that faithfully recorded salinity of the bay waters and was abundant over the past several thousand years, *Elphidium excavatum*. We carefully separated twenty to thirty of the tiny fossilized shells from successive layers in the core for oxygen isotope analysis on the mass spectrometer in our laboratory. When plotted against the shells' depths and ages in the core, our measurements revealed that salinity in the bay began to shift from less saline waters to more saline conditions about 1,800 years ago, peaking in salinity about 600 years ago. These results corroborate those from Suisun Marsh—that dry conditions in California persisted for at least a thousand years (see figure 24).

Paleoclimate research throughout California has produced evidence that, taken together, adds up to some sobering results. In one study, sediments accumulating in an ancient lake located west of the Sierra Nevada revealed major periodic swings in precipitation. In the late 1980s, Scott Stine, then a graduate student in the Geography Department at the University of California, Berkeley, studied the geomorphic and sedimentary history of Mono Lake, a lake that has no natural outlet.

Stine was able to study the exposed delta sediments along the mouths of the streams that feed into the lake because these deposits were exposed between 1941 and 1990 when the inflowing tributaries were diverted to southern cities, causing the lake level to drop. More deposits were exposed during the drought of the late 1980s. Stine analyzed and radiocarbon dated the sedimentary sequences, delineating patterns of alternately higher and lower lake levels over the past 4,000 years. His results showed that, in the past 2,000 years, Mono Lake experienced an extended low stand that began about 1,600 years ago and remained relatively low, dropping to an even lower level between 1,050 and 550 years ago (see figure 20 in the preceding chapter).

Stine also discovered large tree stumps submerged in Mono Lake that were exposed when the lake's surface dropped during the drought of the 1980s. These submerged tree stumps indicate that, at one time, the lake was so small that its shoreline was several tens of feet lower than the present level, allowing the trees now underwater to grow on dry ground. Just like the submerged tree stumps in Lake Tahoe (discussed in chapter 7), these stumps provide strong evidence for extended drought in the past.

A decade later, in the early 1990s, Stine discovered similar tree stumps in lakes, marshes, and rivers throughout the central and eastern Sierra Nevada, including Tenaya Lake in Yosemite National Park, Osgood Swamp, and the Walker River (see figure 25). These trees, mostly Jeffrey pines, are not adapted to soggy soil conditions, but the size of these stumps indicated that they had lived well over 100 years. Thus, they provided clear evidence that, at the time the trees were living, the level of Tenaya Lake must have been much lower than in recent times. Based on the amount the lake level must have dropped for these submerged trees to have grown on dry ground, away from the banks of lakes or rivers, Stine calculated that average annual river flows in the region were only 40 to 60 percent of what they were in the late twentieth century.

FIGURE 25. An ancient tree stump submerged in the West Walker River, eastern Sierra Nevada. (Photo courtesy of D.J. DePaolo, University of California, Berkeley.)

By counting the annual growth rings of these submerged trees, Stine showed that the trees had lived upward of 160 years. The droughts must have lasted at least that long to allow the trees to survive in that location. The trees were finally killed when the droughts ended abruptly, followed by dramatically high rainfall that rapidly raised the lake levels to drown them. Stine dated the outer growth layers of the tree stumps in Mono Lake and in other lakes and rivers in the central and eastern Sierra with radiocarbon methods. He found that the ages of all these trees clustered around two distinct periods: AD 900–1100 and AD 1200–1350.

Tree-ring studies from a broad region of North America indicate that climate conditions over the past 2,000 years became steadily more variable, with especially devastating droughts during this period as compared with earlier times. These records also suggest that the driest period in the West occurred between AD 900 and 1400. For instance, researcher Edward Cook and his colleagues at the Lamont Doherty Earth Observatory Tree Ring Laboratory in Palisades, New York, have studied tree-ring records that reflect the summer values of the Palmer Drought Severity Index—a measure of the moisture content of soil in the root zone. They compared tree-ring records from throughout North America and found that over half of the American

West suffered severe drought between AD 1021 and 1051 as well as from AD 1130 to 1170, AD 1240 to 1265, and AD 1360 to 1382.

Medieval Climate Extremes

The Medieval drought included not only severe drought, as detected in the San Francisco Bay marsh records, tree rings, and tree stumps from Mono Lake and elsewhere in the Sierra, but also severe floods. In fact, several studies suggest that conditions were exceptionally wet in the interval between the two Medieval megadroughts (850 to 750 years ago). For instance, Carson Sink lakes in the Carson Desert of the Great Basin of Nevada reached their highest levels between 915 and 650 years ago. Using water balance models, researchers at the Desert Research Institute calculated that these lakes would have required four times the modern stream inflow for 70–80 years in order to reach such high stands. More evidence comes from the White Mountains of eastern California, where bristlecone pine growth rings indicate this to be the wettest period in the past 1,000 years. Tree rings from the southern Sierra Nevada also suggest exceptionally wet conditions during this period between the droughts.

Extremes in wet and dry during the Medieval period were not the only anomalies. The trees of the Sierra Nevada tell us that, for at least the latter part of the period, conditions were also warmer than average. Connie Millar, a research scientist for the U.S. Department of Agriculture, discovered the remains of trees that had been buried by the ash of a nearby volcanic eruption in the high peaks of the eastern Sierra Nevada 650 years ago. Several species of these trees were growing much higher in the mountains than they do today, suggesting that temperatures in eastern California were more than 3°C (5.4°F) warmer toward the end of the Medieval drought than today.

Millar's results were supported by those of dendrochronologist and paleoecologist Lisa Graumlich when she was the director of the Laboratory for Tree-Ring Research at the University of Arizona. Analyzing tree rings from foxtail pines and western junipers in the southern Sierra Nevada, she found that the period between 900 and 625 years ago exceeded twentieth-century warmth.

Other scientists from the University of Arizona, including Connie Woodhouse and David Meko, studied trees in the Colorado River Basin. Their tree-ring data indicated a particularly warm and dry period during

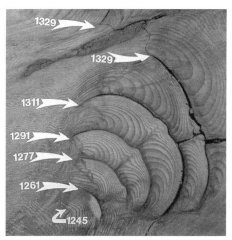

FIGURE 26. (A) Fire scars at the base of a mature Giant Sequoia trunk from the Sierra Nevada (http://commons.wikimedia.org/wiki/File:Fire_scars_in_sequoia_trunk). (B) A cut section of a Giant Sequoia trunk from Tuolumne Grove, Yosemite National Park, California, showing AD dates of fires (photo courtesy of Thomas Swetnam, Laboratory of Tree-Ring Research, University of Arizona).

the mid-twelfth century, when average temperatures were 1.8°F above the long-term average and Colorado River flows were 15 percent below normal.

Increased Wildfires

As conditions grew warmer and drier during the Medieval drought, wildfires would be expected to be larger and more frequent. Very large and old trees can provide a long history of past forest fires. Giant Sequoias (*Sequoiandendron giganteum*), the massive redwoods growing in about seventy-five distinct groves along the mid-elevations of the western Sierra Nevada, are among the most useful of such living archives. These spectacular trees can live up to 3,200 years or more and can exceed 250 feet in height and 35 feet in diameter. They can withstand numerous forest fires throughout their lives, and these fires leave behind scars (see figure 26). Thomas Swetnam, the current co-director of the Laboratory of Tree-Ring Research at the University of Arizona, discovered that these fire scars are visible on their annual growth rings and can therefore be used to assess the frequency of past fires.

Swetnam and his team sampled Giant Sequoias from five groves between Yosemite National Park and Sequoia National Park, far enough apart that

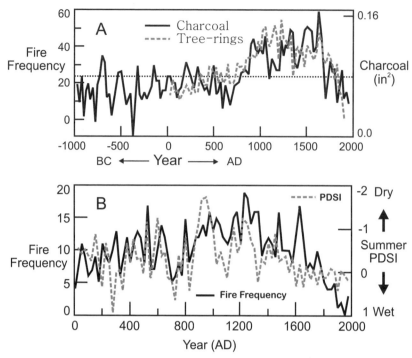

FIGURE 27. (A) Fire frequency in the Sierra Nevada based on charcoal abundances from sediment cores from five meadow or bog locations, compared with fire histories based on fire scars from Giant Sequoias. Note that fire frequency was lower prior to 2,300 years ago (300 BC) and particularly high during the Medieval period (AD 900–1600). Multidecadal- to centennial-scale fluctuations in fire frequency are also exhibited in the record. (B) Fire frequencies (solid line) and means of the Palmer Drought Severity Index (PDSI, dashed line) showing the close correlation between the two. Fire frequencies, mean summer temperatures, and PDSI all peaked at the end of the Medieval period and declined throughout the Little Ice Age. (Figure redrawn from Swetnam et al. 2009.)

individual fires could not have spread from one grove to the next. Using a chainsaw, they sampled cross sections inside the fire-scar cavities at the base of dead trees. The team dated the trees using ring-width patterns and recorded the fire scars contained within annual rings. Their analysis reveals that fires were more frequent from AD 900 to 1400—coeval with the Medieval Climate Anomaly, when an average of thirty-six fires burned every century. During the centuries preceding the Medieval period and immediately following it, the fire frequency was substantially lower, with an average of twenty-one fires per century (see figure 27A). These results are consistent with sedimentary char- coal analyses from Sierra Nevada meadow cores, which also show a marked increase in fire frequency during the Medieval drought (see figure 27B).

Archaeologists and climate researchers are increasingly often teaming up to understand the environmental conditions during key periods of past climate or environmental change. The environmental conditions can provide context and breadth for their understanding of the lives and fates of the ancient inhabitants of the West.

One such collaboration has been in the Four Corners region of the Southwest, where archaeologists have long observed the so-called "Anasazi collapse." Tree-ring studies from this region show unusually narrow rings during the periods AD 1020–1050, AD 1130–1180, and AD 1276–1299. Each of these series of narrow rings reflects periods of slow growth, indicating drought, which corresponds to dramatic population declines in the Southwest as interpreted from archaeological evidence. For example, researchers studying the Medieval drought in New Mexico have produced tree-ring records that focus on summer rainfall for the region spanning the past 1,200 years. Their results reveal that the droughts from AD 1130 to 1180 and AD 1276 to 1299 were tied to failures of the summer monsoon rains.

A similar result was found in a pollen study from southwestern Colorado that focused on the prevalence of piñon pines, a common species in the area that depends on summer monsoon rains for seedling germination. The pollen record reveals a sharp reduction in seedling success in the region lasting over 400 years, from AD 1250 to 1650, indicating more failure of summer monsoon rains and less seedling establishment.

Larry Benson and his colleagues at the U.S. Geological Survey have recently examined the possible impact of this failure of the summer monsoons on the Ancestral Pueblo by comparing many proxy records of climate from throughout the region—including the tree-ring records in New Mexico and the pollen study from Colorado—to the archaeological records of Ancestral Pueblo migrations. Benson found that major human migrations occurred during prolonged periods of drought, particularly when the summer monsoons failed. To better understand how the droughts would have affected these populations, the team applied a model of maize production based on Zuni and Hopi agriculture as the closest approximation of the early Ancestral Pueblo agricultural practices. In this region, the crops were planted in the spring and depended on summer rains for water. The models show that, during multiple years of summer drought, annual harvests of maize would not have been sufficient to sustain human populations. Most likely, a good harvest would have yielded enough food for the current year, plus one year's reserve, but no more. Multiyear failed harvests would have been devastating to the growing population.

There is evidence that human populations were feeling the impacts of these droughts throughout the Southwest. The archaeological record shows that populations at Mesa Verde in southwestern Colorado began building dams and canals, presumably to store and then transfer water onto terraced fields as a response to the decreased reliability of rainfall. The region was occupied between about AD 550 and 1275 before it was suddenly abandoned.

In Utah and western Colorado, the Fremont people vanished abruptly at the same time as the Ancestral Pueblo. In many cases, the populations fled to high cliff dwellings. Some of the evidence shows that as crops, particularly maize, began to fail, the people hid their limited food in storage chambers carved into the cliffs. These may have been last-ditch efforts to stave off collapse. The Sinagua culture in central Arizona also abandoned sites throughout the region following an occupation from AD 1125 to 1300.

THE IMPACT ON NATIVE CALIFORNIANS

The unusually dry climate during the Medieval drought appeared to have tested the endurance and coping strategies of even well-adapted native populations in California. Skeletal remains—bones and teeth—are used as indicators of the health and condition of these populations. They show that life in the interior of California was particularly difficult because the drought severely reduced sources of food (nuts, acorns, plants, deer, and other game). Settlements along rivers were abandoned, and trade between inland and coastal groups broke down.

As water supplies dried up, conflicts—even battles—between groups arose over territory and resources. Populations had grown in better times prior to the drought, leading to more territorial disputes and competition over oak groves. Along the Southern California coast and on the Santa Barbara Channel Islands, the Chumash Indians were then too populous to move during the drought as they had during earlier periods of dry climate. Skeletal evidence shows dramatic increases in malnutrition and diseases as well as projectile injuries and head wounds, interpreted as having been inflicted during warfare. This evidence, uncovered by anthropologists Patricia Lambert and Phillip Walker, indicates that deteriorating environmental conditions and dwindling food and water resources led to conflicts between the growing population.

On the whole, coastal groups appear to have fared better than inland groups because most of their food came from the ocean. Increased upwelling of nutrient-rich waters along the coast during this time increased marine productivity and expanded resources such as fisher and shellfish.

Freshwater was scarce, however, and all along California's coast, including San Francisco Bay, sites were abandoned as the desperate populations wandered in search of new water sources. Kent Lightfoot and Ed Luby have speculated that coastal groups began moving over larger territories as resources became scarcer, perhaps alternating between inland and coastal sites on a seasonal basis. The mobility of these populations allowed them to endure this difficult period.

THE MEDIEVAL CLIMATE ANOMALY:
A GLOBAL PHENOMENON?

Across the globe, the archaeological record contains examples of societies rising or falling during the Medieval period. The Ancestral Pueblo culture did not survive these changes, as we have discussed above, whereas others thrived. The fate of each culture seems to have been tied to its unique local situation. But the fact that so many regions experienced upheavals in the same period raises questions about larger-scale external influences during this period of history.

Paleoclimate studies throughout the Northern Hemisphere continue to fill in the details of regional climate from the ninth to the fourteenth centuries. While societies in the American Southwest were suffering catastrophic decline caused by drought, societies in northern Europe, where average temperatures were about 1.8°F warmer than today, were flourishing. This seemingly small increase in temperature was sufficient to unleash major historic events. Eric the Red was leading Norse settlers to Iceland and Greenland to establish colonies, and European farmers were expanding into new lands previously too cold to support agriculture—even planting vineyards in England. This era featured the growth of towns and city-states; science and art blossomed; trade expanded; and communication improved among countries. In

Europe, the changes in climate fed the blossoming of societies. But in the American West, the changes in climate brought devastation.

We now see that the Medieval period, initially believed to be a worldwide period of unusual warmth, was not a uniform experience across the globe. In a seminal 1994 review of the topic, appropriately entitled "Was there a 'Medieval Warm Period' and if so, where and when?," Malcolm Hughes and Henry Diaz came to the conclusion that the 500-year period of the Middle Ages was a time when various regions across the world experienced episodes of warmer temperatures, but not all at the same time. Furthermore, in some regions, including the southeastern United States, the Mediterranean, and parts of South America, there appeared to be very little change in average temperatures. The authors proposed that the misnomer "Medieval Warm Period" be replaced by the term "Medieval Climate Anomaly." In this book, too, we will use the latter term in the remainder of our chapters.

The article was particularly timely for California and the West, since the 1980s and 1990s were marked by prolonged droughts alternating with extremely wet years with damaging floods. Many people, including water managers, were becoming increasingly interested in understanding the region's climate. In this context, information about the lake-level and tree-stump history of Mono Lake and the increased salinities in San Francisco Bay provided stark warnings of how climate could affect our lives in the future.

BACK TO THE MARSHES

We return to the Suisun Marsh record because it shows that a long period, including the Medieval Climate Anomaly, was clearly drier in San Francisco Bay's vast watershed than it was before or after. However, the marsh record lacks any evidence of the two "megadroughts" detected in the central to southern Sierra. To further understand the changes in the bay estuary's watershed during these droughts, our research group took a second look at the San Francisco Bay marsh sediments, this time focusing on the clays and silts that wash into the bay from its watershed. These sediments carry mineral and chemical fingerprints of their origins, such as the volcanic rocks (mainly andesites) of the northern Sierra Nevada range and southern Cascades, or the granites and granodiorites of the central and southern Sierra. As these sediments are carried into San Francisco Bay, the largest particles settle to the bottom of the estuary while the finer, lighter ones are washed onto the surrounding marshes, forming steadily accumulating layers.

The Sacramento River drains the northern part of the watershed, carrying sediments from the northern Sierra Nevada and southern Cascades, and the San Joaquin River drains the southern part of the watershed. The mineral composition of the sediments is as valuable to paleoclimatologists as fingerprints are to police detectives, providing evidence of origin. The changing proportions of sediments from the northern and southern parts of the watershed provide details of how climate in Northern and Southern California has varied over time, and these changes have shown that the Medieval droughts were indeed felt throughout the watershed of the estuary, in both northern and southern portions—but not at the same time.

During the first of the megadroughts, which began around AD 900, the southern Sierra Nevada experienced the initial severe conditions. Run-off flows were so low that almost no sediments from that region reached the bay. In the northern portion of the watershed, conditions remained normal at first, and the Sacramento River flows continued to transport sediments. However, as the drought persisted, it spread northward, causing reduced flows of the Sacramento River and therefore bringing less sediment from that region to the bay. At the same time, flows from the San Joaquin River, draining the southern watershed, increased somewhat (around AD 1080). The second major drought (between AD 1200 and 1400) shows a similar pattern.

When we compare these results with tree-ring studies from the northern Sacramento River watershed and the southern San Joaquin River watershed, we see agreement: the droughts across California did not always occur at the same time, but alternated between the northern half and the southern half of the state. The Medieval Climate Anomaly was a period of unusual climate, but the impacts on precipitation over the American West were not uniform in magnitude, duration, or geographic location. What could have caused these unusually dry periods in the West?

CLUES FROM THE PACIFIC OCEAN

The search for the cause of this period of variability and climate extremes led researchers to turn their attention to the primary source of the American West's climate: the Pacific Ocean. Some of the first clues emerged from the ocean just off coastal Southern California.

Sediments on the coastal ocean floor contain the remains of single-celled organisms that form calcite shells, called foraminifera (or "forams"), which contain information about oceanic conditions like sea surface temperature. Off the coast of Santa Barbara, these forams have been used to assess sea surface temperatures over the past ten thousand years using the oxygen isotopic composition of a particular species of foram, *Globigerina bulloides*. This species lives exclusively in surface waters, so its shell incorporates oxygen directly from the surface waters and faithfully records changes in ambient sea surface temperature. Analysis of these shells from cores broadly coincident with the Medieval period has shown that the surface water temperatures were 2–3°F colder from AD 700 to 1300 than they are today. Cooler than average waters in the eastern Pacific are often associated with La Niña conditions, which bring drought to the American Southwest, but more evidence from the tropical Pacific Ocean would be needed to confirm the connection during Medieval times. We will discuss this and other evidence for the ocean-atmosphere changes that occurred during this time in chapter 11.

In light of the evidence for a warm, dry period in the West between about 1,800 and 550 years ago, with particularly deep droughts toward the end of that period (from 1,050 to 550 years ago), current inhabitants of the western United States face sobering questions: How common are such severe and prolonged droughts? Do they represent the most extreme events we can expect? When will the next one strike? Will they be more frequent and severe as the region warms in the coming decades? How would they affect society today?

And what of the opposite climate extreme—the deluges and floods that ended the droughts? How often do these floods, with magnitudes similar to the 1861–62 event or even greater, recur? To further explore these questions, we turn next to the Little Ice Age. This cool period, which followed the Medieval Climate Anomaly and lasted until the nineteenth century, brought several megafloods of a magnitude that no one alive today has yet experienced.

TEN

The Little Ice Age

MEGAFLOODS AND CLIMATE SWINGS

The lessons of the Little Ice Age are twofold. First, climate change does not come in gentle, easy stages. It comes in sudden shifts from one regime to another—shifts whose causes are unknown to us and whose direction is beyond our control. Second, climate will have its sway in human events. Its influence may be profound, occasionally even decisive. The Little Ice Age is a chronicle of human vulnerability in the face of sudden climate change. In our own ways, despite our air-conditioned cars and computer-controlled irrigation systems, we are no less vulnerable today. There is no doubt that we will adapt again, or that the price, as always, will be high.

BRIAN FAGAN, *The Little Ice Age*

SAN FRANCISCO BAY

THE DEEP DROUGHTS OF THE Medieval Climate Anomaly eventually drew to a close around AD 1400. For much of the 150 years prior to this date, extreme flooding was relatively rare. In California, people who lived in the broad Central Valley and around the San Francisco Bay would have seen irregular rainfall for hundreds of years, with the winters often failing to deliver the big storms that filled the lakes and fed the mountain snowpack. Around the bay, the mounded villages were empty most years; since the drought had begun, the creeks flowing to the marshlands dried early in the summer, forcing the people to make long treks for freshwater or, if their camps were inland near the larger creeks, long foraging expeditions to the mudflats in the bay to collect mussels and other shellfish. The native populations throughout the region suffered, some perished, and some managed to adapt to the dry conditions.

The pattern of weather began to change, however, and those who paid attention would have noticed the series of cold storms that now began the wet season, dumping rain on the coastal settlements and throughout the Central Valley and the lower mountain slopes. Winters were bringing more storms that were colder, and in the Sierra Nevada the snowpack grew thicker and lower on the mountain slopes than it had been in human memory. This period was known as the Little Ice Age, and, elsewhere around the globe, rivers and canals froze throughout Europe, and valley glaciers grew larger in northern and central Europe, New Zealand, and Alaska. The Thames River in London froze at least eleven times in the seventeenth century. Lower summer temperatures throughout Europe decreased the growing season by several weeks, causing crop failures in Scotland, Norway, and Switzerland and leading to widespread famine. In Norway, there is evidence for more frequent landslides, avalanches, and floods, and, in the Alps, expanding glaciers occasionally crossed valley floors, damming streams and forming glacial lakes. These lakes would on occasion break through the ice dams holding them back, flooding the valleys below.

In California, some of the storms originated in tropical regions of the Pacific, delivering copious amounts of warm rain that led to flooding. These storms would have swept over the Coast Ranges, across the Central Valley, and up into the Sierra Nevada, delivering heavy rain. The rain and the rapidly melting snow would have caused small creeks and rivers to swell into raging torrents, ripping up vegetation and soils in the mountains, and turning California's Central Valley into a vast inland sea.

Many proxy climate records from throughout California contain evidence that such extreme winter floods occurred between 900 and 150 years ago. Tidal marshes around the San Francisco Bay contain buried evidence of these events in their sediments. Normally, the inflowing river waters spread across the marshes, depositing only a small amount of the finest sediments—clays and silts. Floodwaters, however, carry larger particles in their higher energy flows, depositing a layer of sand and silt on the marshes. Marsh sediment cores reveal that one such layer was deposited on the marshes around AD 1100, and two others around AD 1400 and AD 1650.

Sediment cores taken from beneath the San Francisco Bay itself provide more evidence of the flooding: a gap or hiatus in the sedimentary sequence suggests erosion of hundreds of years' worth of accumulated sediments from the bay floor. The date for this hiatus matches the earliest of the sand layers in the marsh cores. That such an event is recorded in San Francisco Bay estuarine and marsh sediments suggests a regionally significant flood, affecting almost half of the state—the drainage area of the bay.

Though considered "young" in geological terms, these large floods were not documented by humans living at the time and left no evidence in the archaeological record. To understand them from our standpoint in the twenty-first century requires the same kind of detective work described in earlier chapters to uncover older climate patterns. Below, we describe several key examples of evidence for megafloods throughout the state of California.

SOUTHERN CALIFORNIA FLOOD EVIDENCE

Some of the most compelling evidence of large floods in California has been found off the southern coast, once again in the favorable environment of the Santa Barbara Basin. The location and conditions of that basin have combined to produce one of the most detailed and complete records of climate and environmental change anywhere in the world's coastal oceans, including a record of runoff into this coastal environment during extreme storms.

Arndt Schimmelmann, a marine geochemist at Indiana University, has analyzed the environmental history contained within the Santa Barbara Basin's deep sediments. He first looked at X-ray images of sediment cores, which elucidate differences in the density of the sediments; annual sediment layers show clearly as alternating light and dark stripes. Schimmelmann found that some layers were much thicker than others. When the core was cut open, he found that the thicker layers were olive-grey, and, under the microscope, he could see that these olive-grey grains included tiny angular lithic fragments. These particular sediments were probably eroded off the hill slopes in Southern California and subsequently washed into local rivers, which transported them to the ocean floor. Schimmelmann concluded that these sediment deposits must have resulted from enormous flood events—"megafloods," as he called them.

Schimmelmann was intrigued by what he saw in the Santa Barbara Basin sediments, and he began looking for evidence of flood events in other records of past environmental and climate change that occurred at the same time. He found that other researchers in the coastal region had described evidence of large floods contained in sediments from lakebeds, floodplains, and submarine basins along the coastline. These are typical settings for finding traces of ancient floods. As floodwaters rage down slopes and across the landscape, they scour the hills, picking up clay, silt, sand, and even gravel and carrying them entrained in the swollen current. Eventually the velocity of the rivers

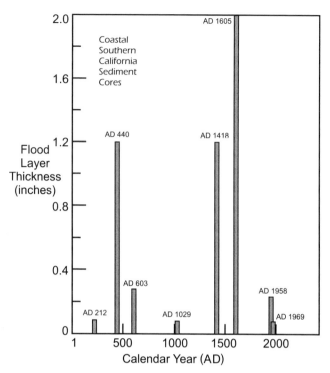

FIGURE 28. Flood sediment layers for the past 2,000 years from Santa Barbara Basin cores. (Figure courtesy of Arndt Schimmelmann, University of Indiana, redrawn by B. Lynn Ingram.)

decreases upon reaching sea level, and the rivers release their burden: first the larger gravels, then the sands, and finally the silts and clays. Nature rebuilds after these events, and in time the flood deposits are themselves buried beneath other sediments.

Ultimately, Schimmelmann found evidence for six megafloods over the past 2,000 years in Santa Barbara Basin sediments that would have affected the entire region. The flood layers could be precisely dated because the sediments are composed of annual layers, analogous to the growth rings in trees. Based on the ages of these megafloods, they appear to have recurred on average every 200 years: AD 212, 440, 603, 1029, 1418, and 1605. During the period straddling the Medieval Climate Anomaly, two cycles were skipped, with 400 years after the floods that occurred in AD 603 and AD 1029.

Schimmelmann reasoned that the thickness of the flood layers is proportional to the size of the event. The thickest layers were deposited before the Medieval Climate Anomaly (AD 212, 440, and 603) and during the Little

Ice Age (AD 1418 and 1605). The thinnest flood layer occurred during the Medieval Climate Anomaly, in AD 1029 (see figure 28). These results suggest that periodic megafloods may be a normal part of the larger cycle of climate in this region, and the Medieval Climate Anomaly may have been anomalous in part for its absence of such floods.

The largest flood to hit the Santa Barbara Basin over the past two millennia occurred in AD 1605, leaving a sediment layer two inches thick. Almost as large, the AD 440 and 1418 floods each had a thickness of one and a quarter inches. For perspective, consider that the two major floods that struck the region during the twentieth century—the floods of 1958 and 1969—left sediment layers that were only 0.24 and 0.08 inches thick, respectively. In its entire history as a state, California has yet to experience one of the megafloods that Schimmelmann found to be a repeating part of the region's natural climate. Schimmelmann points out that the periodicity reflected in his record suggests the region is due for another one soon; it has been more than four hundred years since the last megaflood, which occurred in AD 1605.

THE MOJAVE DESERT UNDER WATER

For most of the Holocene, the Mojave Desert of southeastern California has remained one of the driest places in the West, with annual rainfall less than four inches per year in recent times. Only during the wettest years of the twentieth century (every fifteen to thirty years or so) did small ephemeral lakes form in the region, lasting just two to eighteen months before drying up. However, paleoclimate researchers have examined sediment cores extracted from a playa located in the Mojave and found evidence of past periods that were wet enough to form much larger and long-lived lakes. This playa, called Silver Lake, lies at the terminus of the Mojave River. During the late Pleistocene, Silver Lake was much larger but has remained mostly dry since about 8,000 years ago. Only during two periods since the late Pleistocene were conditions wet enough to create a lake that lasted several decades.

Cores taken from the Silver Lake playa reveal two layers of clay sediments deposited within the lake. Within those cores, the lake sediment layers lie between sediments deposited by the Mojave River on the desert floor. Radiocarbon dating of the lake sediments reveals two extremely wet periods in the past: the Neoglacial (about 3,600 years ago), and the Little Ice Age (about 400 years ago). A period of cool summer temperatures, with a greater than average number of storms per year—for decades to up to a century—would have been required to form these lakes in the Mojave.

The ominous history of ancient megafloods identified in Southern California is not unique. Researchers in the northern half of the state have found deposits suggesting floods of similar magnitude and frequency—from river valleys to the higher mountain ranges. One of these studies was located on the floodplain of the Sacramento River, the largest river in California and the one that drains much of the northern half of the state. Geographers Roger Byrne and Don Sullivan from the University of California, Berkeley, have found evidence for large floods in sediments from beneath oxbow (or arc-shaped) lakes on the Sacramento River floodplain of California's Central Valley. They were not initially looking for evidence of megafloods; rather, they were evaluating how the burgeoning population in California after the Gold Rush had influenced the natural vegetation in the region. But as they examined their sediment cores, they found an unexpected legacy of major flood events.

To understand what Sullivan and Byrne discovered, it helps to visualize the setting of their research. Floodplains, by their nature, contain the most accurate picture of past flooding events. During a flood, the waters swell out of their channels and spread across the surrounding plain, depositing layers of sediment. These layers build on each other, flood after flood, over hundreds to thousands of years. The thickness of the layers yields information about the size of the floods. The lakes within the floodplain, though some are short-lived, contain records that can be interpreted both for past flood events and for the climate conditions under which the lakes formed.

The lake Sullivan and Byrne chose to core, Little Packer, began forming 800 years ago. Once a bend in the Sacramento River, Little Packer was abandoned by the meandering main channel and was left behind as an oxbow lake. This occasionally occurs when rivers meander in wide and gently sloping floodplains, as the Sacramento does in the Central Valley. Such abandoned lakes become recorders of large flood events that temporarily reunite them with their rivers. Sediment-laden floodwaters spill into them, and the sediment settles to the bottom, forming a distinct layer within the normal lake sediments.

Sullivan and Byrne took a series of sediment cores from beneath Little Packer Lake and brought them back to Byrne's laboratory. During the routine X-ray analyses, they noticed some unusual bright layers in the cores that turned out to be rich in river-borne sand and silt. These denser sand layers shone white next to the darker, more organic layers produced by algae growing in the lake during normal (non-flood) years. Sullivan and Byrne reasoned that, since the lake had no direct hydrologic connection to the river except during major floods, these deposits must have occurred when the Sacramento River overtopped its banks and carried suspended sediments into Little Packer Lake. Because larger floods transport more sediment, the thickness of the sand layers provides an approximate measure of the relative size of the flooding event. Sullivan and Byrne also compared the flood layer thicknesses to floods of known magnitude, such as the catastrophic 1861–62 event.

Eager to understand more about how these floods fit into California's natural history, Byrne and Sullivan used radiocarbon analysis to date the cores and estimate the ages of the flood layers. Although these ages lack the precision of ages that are determined by counting annual varves in the Santa Barbara Basin, they can still provide an approximate measure of the flood frequencies. The researchers calculated that, over the past 800 years, floods at least the size of the 1861–62 event struck Northern California on average every 100 to 120 years.

This is bad news to state water planners, who have estimated that the 1861–62 flood was a "500- to 1,000-year event," which was determined by plotting known flood magnitudes against their recurrence intervals from California's Central Valley over the past century. This so-called flood frequency curve then allows hydrologists to extrapolate the recurrence intervals of much larger floods (like the 1861–62 event) using their estimated magnitudes. Because paleoflood records provide actual data about the magnitudes and recurrence intervals, they should be considered along with those estimated from the flood frequency curves. For instance, the research of Byrne and Sullivan revealed evidence of four even larger floods that dwarfed the 1861–62 event over the past 800 years. Like Schimmelmann with his results from the Santa Barbara Basin sediments, Byrne and Sullivan found evidence of megafloods recurring in the region approximately every 200 years.

Evidence of megafloods has also emerged from the far northwest corner of California. In that mountainous region, floods ravage the steep, narrow canyons in the Klamath and northern Coast Ranges, where flood debris tends to form thicker, coarser layers than are found in the broader and more gently sloping floodplains of the Central Valley. In those mountains, only the larger flood deposits are preserved, in contrast to the sequence of flood deposits that built up over time in the Sacramento Valley. This is due to the erosive power of these floods in the narrow canyons that destroys any evidence of older, smaller floods. Researchers at the U.S. Geological Survey discovered two enormous flood deposits preserved in this region, estimated to be larger than one of the most catastrophic historical floods in Northern California, which occurred in 1964. These two floods were dated by counting the growth rings of the oldest trees rooted in the flood deposits, and they occurred in approximately AD 1600 and 1750. A flood layer dated between AD 1750 and 1770 was also found in the Sacramento Valley by Byrne and Sullivan, and the AD 1600 flood deposit was likely the same as the AD 1605 events found in both the Sacramento Valley and the Santa Barbara Basin. The evidence for a 200-year recurrence interval for floods as large or larger than the 1861–62 event in California is growing, bringing more urgency to flood policy and planning efforts in the state.

THE AD 1605 MEGAFLOOD

The evidence for a megaflood occurring in AD 1605 has been found throughout California, including in the northwest, central, and southern regions as well as in San Francisco Bay, making this a truly extraordinary event even by "megaflood" standards. In the Little Packer Lake record from the Sacramento River basin, this flood was estimated to be at least 50 percent greater than any of the four 200-year events that Sullivan and Byrne detected in their cores. It must have been a catastrophe of gargantuan proportions, clearing hill slopes, drowning the Central Valley, and inundating most of the state before ultimately flowing through the Sacramento–San Joaquin Delta, through the San Francisco Bay, and out to the Pacific Ocean.

Sediment cores taken from beneath the southern San Francisco Bay contain oxygen isotope evidence that this time period was marked by very low salinity in the estuary, presumably from the high river inflows from the watershed. In addition, cores taken from the northern reach of the bay, closer to where rivers flow in through the delta, contain a sedimentary hiatus, suggesting that the flood was large enough to erode and wash away the sediments out to the Pacific Ocean.

Intrigued by the phenomenon of cataclysmic floods and their possible causes, Schimmelmann combed through the scientific literature and compiled a list of other global events that occurred at the same time as these megafloods. He found that, around AD 1605, at least two major volcanic eruptions occurred: the Huaynaputina volcano in Peru in 1600, and other volcanic evidence revealed in an Antarctic ice core. He also discovered numerous paleoclimate records worldwide showing that the years between 1600 and 1610 were unusually cold, particularly during the summer, with the summer of 1601 being the coldest in the Northern Hemisphere of the past millennium. These records were largely based on tree-ring records from California, Idaho, the Canadian Rockies, Subarctic Quebec, the Gulf of Alaska, New Zealand, Mongolia, and Norway. Ice-core records from both Greenland and Antarctica also suggest unusually cold conditions. But was there a connection between the volcanic eruptions, Northern Hemisphere cooling, and the California megafloods? We will return to this question in the next chapter, where we discuss possible causes of past climate change.

THE LITTLE ICE AGE AND THE WEST

These megafloods occurred during the Little Ice Age, an event that, as described above, left traces throughout the Northern hemisphere between the fifteenth and nineteenth centuries. In the western United States, glaciers also advanced in the Sierra Nevada, Cascades, and Rockies. Conditions became wetter in the West, and floods were larger and more frequent, as discussed above. In the Sierra Nevada, abundant moisture allowed trees to grow faster, putting on wider rings. East of the Sierra Nevada, Mono Lake rose to its highest level in centuries. In the Great Basin, Owens and Pyramid lakes were also high.

Forest fires in California were reduced during the Little Ice Age. For instance, though the Giant Sequoias in the southern Sierra Nevada bear scars from fires over the past millennia, those scar records show a marked decline in fire frequency after AD 1600. The records also closely track the summer values of the Palmer Drought Severity Index (PDSI), a measure of the moisture content of soil in the root zone. Tree-ring records contain a history of the PDSI. A comparison between the PDSI and the fire-scar records shows that three-fourths of the largest fire years occurred during the driest years. Mean summer temperatures, the PDSI, and fire frequencies all peaked about AD 1200 during the Medieval period, followed by a steady decline in these factors (suggesting a cooler, wetter climate with fewer fires) during the Little Ice Age (see figure 27B).

Additional evidence for increased wetness during the Little Ice Age has been found downstream of the Sierra Nevada, in the Central Valley. Tulare Lake, filled with increased river inflows and floodwaters, reached a high stand during the Little Ice Age. San Francisco Bay became fresher as river inflows increased. Vegetation records from the marsh sediment cores surrounding San Francisco Bay reveal a shift to a dominance of freshwater species, reflecting greater inflows from rivers swollen with heavy rainfall in the watershed.

CLIMATE SWINGS, WILDFIRES, AND DROUGHTS

Many of the paleoclimate records show increased variability during the Little Ice Age, with a generally wetter and cooler climate punctuated by episodes of drought—sometimes severe. Droughts have been detected in tree rings from the Sierra Nevada, salinity levels in San Francisco Bay, and lake levels of Mono, Pyramid, and Owens lakes. At Mono Lake, for instance, Larry Benson and his colleagues examined sediment cores using the oxygen isotopic ratios of the sedimentary carbonates to determine changes in the lake's surface level. As discussed in chapters 5 and 7, when the inflow of water entering a closed-basin lake exceeds the rate of evaporation, the lake expands and the ratio of oxygen-18 to oxygen-16 is lower. When evaporation is the dominant process, however, the lake shrinks, and the lighter oxygen-16 water molecules evaporate more readily, leaving behind more oxygen-18 molecules in the lake. Benson and his colleagues reconstructed the relative changes in lake level based on the oxygen isotopic ratios in carbonates, and they have shown that the water level in Mono Lake dropped five times over the past 300 years, implying severe droughts. These dry periods lasted for several years, causing a significant drop in lake levels, centered on the dates AD 1710, 1770, 1820, 1850, and 1930.

The San Francisco Bay records also contain evidence of these periodic swings in climate seen in the Sierra during the Little Ice Age. Oxygen isotope measurements of sediments from the northern part of the estuary have enabled the authors to reconstruct the bay's salinity (as discussed in chapters 5, 8, and 9). The results reveal multiple fluctuations in salinity over the centuries, exhibiting 55-, 90-, and 200-year cycles.

In the American Southwest, climate during the Little Ice Age appears to have been highly variable. Tree-ring studies from the Colorado Plateau show that Colorado River flows experienced major fluctuations during that period. Looking at reconstructed 25-year average flow conditions, researchers have found that it was a time of generally cooler conditions, with at least six wet periods marked by significantly high flows. However, there were also periods of significantly low river flows, including AD 1564–1600, AD 1844–48, and AD 1868–92. Over the past 450 years, the long-term annual average flow for the Colorado River was 14.6 million acre-feet. Flows during the identified dry periods were as low as 9.6 million acre-feet, only 66 percent of the long-term average.

Although these flows were low, they were not as low as during the Medieval Climate Anomaly. In particular, from AD 1130 to 1154, Colorado River flows were 85 percent lower than the twentieth-century mean. We will discuss this variability and its origins further in the next chapter.

THE DRY-WET CLIMATE KNOCKOUT

Some of the paleoclimate records reveal repeated patterns of dry and wet climate extremes, such as severe droughts immediately followed by catastrophic floods—or the opposite, wet periods followed by droughts. These combinations can wreak havoc on ecosystems, especially when accompanied by unusually large wildfires, soil erosion, and disease.

One example of a "dry-wet knockout" has been documented by Scott Mensing, Roger Byrne, and Joel Michaelsen, who analyzed the flux of large charcoal particles and pollen deposited in sediments cored from the Santa Barbara Basin in Southern California. Their analysis reveals that the largest fires—those often associated with warm, dry Santa Ana winds blowing from the northeast to the Pacific during the fall months—have repeatedly occurred in the first year of a drought that follows an extended wet period. Moreover, Mensing and his colleagues showed that between twenty and thirty large wildfires occurred over the past 560 years, the majority following unusually wet periods.

Examples of this pattern in the twentieth century include the wildfires of 1955 and 1964, which followed two of the wettest winters of the century. Investigators reason that vegetation grew prolifically during years with greater precipitation. When followed by unusually warm and dry summer conditions, the additional vegetation provided extra fuel for unusually large forest fires.

Periods of deep drought that are followed by heavy rain and flooding can also cause problems. This phenomenon has been observed both in the past century and in the more distant past. Prolonged, multiyear droughts followed by severe storms and massive flooding are nature's version of a one-two climate punch. University of Nevada, Reno, dendroclimatologist Franco Biondi and his colleagues have studied these patterns and dubbed them "dry-wet knockouts" because the floodwaters often wash away hillsides that have been denuded of vegetation during extended drought, transporting huge volumes of sediment to coastal waters. There are many examples of extended droughts—including those that occurred during Medieval times—terminated by extremely wet years, which often lead to a chain-reaction of impacts on local ecosystems.

During multiyear droughts, mountain forests become dangerously dry, and summer lightning strikes spark wildfires that sweep through and engulf the forests. Subsequent heavy rains generate floods that can easily erode fire-charred slopes, washing massive amounts of sediment and charcoal downhill, eventually into streams and lakes. Once on the move, such sediments raise the bed-level of the rivers, further adding to flood potential. These sediment-laden waters can disrupt aquatic ecosystems downstream in ponds, lakes, estuaries, and even the coastal ocean.

Evidence of these past knockout floods can be found throughout the West. Tree stumps found in lakes in the southern and eastern Sierra Nevada are what remain of medieval trees that grew for over a century during the prolonged droughts of the Medieval Climate Anomaly and were suddenly drowned when the drought was terminated by heavy rains and huge floods. Downstream in the San Francisco Bay tidal marshes, sediment cores reveal evidence of large floods dating to around AD 1100, 1400, and 1650. These appear to have occurred during the interval between the Medieval megadroughts and immediately after the Medieval droughts. Similarly, in Yellowstone National Park, Wyoming, evidence of massive debris flows, caused by flash floods, has been dated to the end of the Medieval drought.

The greatest megaflood on record in the West over the past two millennia—the AD 1605 event described above—also followed on the heels of an extended dry period, which occurred at the end of the sixteenth century. And the deep drought during the mid-nineteenth century that peaked in 1860 was followed by the catastrophic 1861–62 flood. More recently, the drought of 1987–92 was followed by the unusually wet winter in 1993, leading to massive mudslides, landslides, and record flooding in some parts of the West.

Because the climate in the West is strongly influenced by the temperature of the eastern Pacific Ocean, paleoceanographers—those who study past ocean conditions—have examined sediment cores from the coastal ocean for evidence of correlations with climate on land. One such researcher, John Barron of the U.S. Geological Survey, used the fossils of diatoms in the Santa Barbara Basin to reconstruct long records of ocean surface water conditions. Barron and his colleagues have shown that, whereas coastal surface water was generally cooler during the Medieval Climate Anomaly than today, the Little Ice Age coastal surface waters tended to be warmer, and more variable, than coastal waters during the Medieval period. Warm waters in the eastern Pacific are associated with El Niño, and, as we will discuss in the next chapter, a growing body of evidence, including coral records from the tropical Pacific Ocean, suggests that El Niño events were stronger and more frequent during the Little Ice Age, leading to wetter and more variable conditions in the West.

The droughts that punctuated this generally wetter period were also shown to be tied to coastal ocean conditions. Larry Benson and his research team compared the dry periods from Mono Lake in the eastern Sierra with sea surface temperature records from coastal California, and they were intrigued to find a close correlation between the droughts and the periods when coastal waters were relatively cool. As we discussed in chapter 4, one of the oceanic cycles that influences climate in the West is the Pacific Decadal Oscillation (PDO). This is an oscillation of water temperature in the North Pacific Ocean, with cooler surface water temperatures associated with the negative phase of the oscillation and warmer surface temperatures associated with the positive phase. Benson and his colleagues found that the drought periods in the Sierra that led to the lowering of the Mono Lake surface level occurred when the PDO was in a negative (cool) phase.

The evidence of past megafloods is a cause of concern for California. Even as we enter a period that is predicted to be warmer and drier, catastrophic flooding will continue to threaten the region. As more precipitation falls as rain instead of snow, flooding during the winter is predicted to inundate the region even more frequently. The deadly climate patterns that have occurred repeatedly in the past are likely to recur in the future.

Climate scientists are currently studying the causes of these past climate changes and weather extremes. Some of these causes are truly remote, such as variations in solar output, and others are close at hand, such as the coal-fired power plants that generate much of our electricity. Needless to say, the so-called anthropogenic, or human-caused, changers of climate are most readily modified—but are also controversial and incompletely understood. To comprehend the full range of climate variability and to better prepare ourselves for an uncertain climatic future, we need to use all the knowledge and tools at our disposal. In the next chapter, a review of some known cycles and oscillations of climate change in the past may suggest the scale on which we can expect such changes to recur.

ELEVEN

Why Climate Changes

CYCLES AND OSCILLATIONS

> Just as the records of past peoples help us understand human
> society, the records of past climates help us learn how the Earth
> system works. And just as modern political scientists can test
> their ideas against the history of humans, Earth system scientists
> can test their models against past climate changes.
>
> RICHARD ALLEY, *The Two-Mile Time Machine*

IN THE PAST SEVERAL CHAPTERS, we have highlighted the work of
paleoclimatologists and the tools that have allowed them to re-create the
features and patterns of ancient climates in the American West. One reason
for their success is the variety of tools and climate archives at their disposal—
from the analyses of tree rings, lake levels, and oxygen isotopes as indicators of
climate wetness and temperature to the interpretation of coarse sediment layers in floodplains, wetlands, and coastal sediments as indicators of past floods.
In this chapter, we explore in more detail the proposed mechanisms that
are responsible for past variations in climate and the persistent tendency of
climate to follow patterns and oscillations.

Over very long timescales (millions of years), climate change is caused
by forces within the earth that drive the very slow movements of tectonic
plates—processes that build mountains and spread ocean basins apart. Here,
we focus on those processes that occur on relatively short timescales (years,
decades, centuries, or millennia). Understanding the causes of climate variations and oscillations over longer time periods gives climate scientists insights
into how the climate works, the full range of climates possible, and what we
may expect in the future.

FIGURE 29. Paleoclimate researchers drilling *Porites* coral heads in the Red Sea using a large hydraulic rig. Coral cores are used to reconstruct past seawater conditions. (Photo courtesy of Konrad Hughen, Woods Hole Oceanographic Institute.)

CORAL AND ANCIENT EL NIÑO

We owe much of our understanding of past climate change on Earth to the tiny polyps that create coral reefs. Each polyp, no bigger than the head of a pin, secretes a calcium carbonate skeleton to protect itself. Collectively, over the millennia, millions of generations of polyps have built the magnificent coral reefs found in shallow tropical seas. Coral ecosystems, among the richest communities of species found anywhere, have persisted for over 200 million years, a span of the earth's history that has seen monumental climate changes. Each generation's fossilized skeletons, stacked upon the last, bear testament to many of those changes.

The most recent changes—those occurring over the past several thousand years of the Holocene epoch—provide valuable information for understanding our current climate. Paleoclimate scientists have developed methods of coring through the calcium carbonate coral skeletons, whose annual layers have recorded information on water temperature, salinity, and nutrients (see figure 29). For instance, scientists have probed coral reefs across the tropical

Pacific Ocean for evidence of the past behavior of the ocean-atmosphere phenomenon known as the El Niño–Southern Oscillation (ENSO). During an El Niño event, the trade winds fail, causing the warm waters in the western Pacific Ocean to shift eastward, warming surface temperatures and shallow coral reefs across the tropical Pacific.

Palmyra Atoll, one of the most spectacular and diverse coral reefs in the world, is located just north of the equator in the central Pacific Ocean, making it ideal for studying past changes in the ENSO. Geochemist Kim Cobb of the Georgia Institute of Technology has been studying coral reefs at Palmyra Atoll for over a decade. She uses scuba gear to dive down to and collect cores from the massive heads of the coral genus *Porites,* which grow up to several meters (6–10 feet) high. Cobb also collects fossilized corals on the island in order to extend the record to older time periods. Once all of these coral cores are taken to her laboratory, they are X-rayed to delineate density changes in their calcium carbonate skeleton. Denser layers are formed during periods of slower growth in the cooler winter months; less dense layers are formed during the warmer summer months, when the coral grows more rapidly. These annual layers, like tree rings, provide a chronology of the coral. Each sample is analyzed for geochemical signals of climate events that took place when the coral was growing.

Cobb and her team have found that the Palmyra corals have faithfully recorded the El Niño events of the twentieth century as reflected in the ratio of oxygen-18 to oxygen-16 in the coral skeletons. Armed with this knowledge, Cobb has explored the frequency and intensity of El Niño over past periods of climatic change. She found that, during El Niño events, the waters surrounding Palmyra become warmer and rainfall increases—both leading to a reduced amount of the heavier isotope (oxygen-18) relative to the lighter isotope (oxygen-16) in the coral skeleton. Conversely, La Niña brings cooler waters and reduced rainfall to the site, leading to a greater proportion of oxygen-18 relative to oxygen-16.

Cobb was able to extend the record back 1,100 years by splicing, drilling, and dating dozens of exposed fossilized coral heads on the beaches of Palmyra. These studies reveal that, during the Medieval Climate Anomaly between the tenth and fourteenth centuries, sea surface temperatures were cooler in the central Pacific than today—conditions associated with La Niña. We know from studying modern climate patterns that cooler waters in the central and eastern Pacific are associated with drier conditions over the American West and northern Mexico. Cobb's results suggest that the Medieval droughts (described in chapter 9), which coincided with the collapse and migration of the Ancestral

Pueblo and other cultures across the Southwest and California, may well have been caused by extended La Niña conditions in the Pacific Ocean.

Another intriguing result of the Palmyra coral record reveals that El Niño events became more frequent and intense during the Little Ice Age—particularly in the seventeenth century—than we have experienced for most of the twentieth century. Today, we know that El Niño events are associated with increased precipitation in the Southwest. As discussed in chapter 10, the cooler climate during the Little Ice Age coincided with wetter conditions and increased flooding in California and the American West.

THE ENSO OVER THE AGES

Studies of the past behavior of the ENSO have led to new understandings elsewhere as well. For example, southern Ecuador is a region adjacent to the tropical Pacific with a climate closely tied to El Niño. Two geologists, Donald Rodbell and the late Geoffrey Seltzer, selected a study site there, Lake Pallcacocha, to assess the effects of the ENSO over the entire Holocene.

El Niño winters typically involve periods of heavy rainfall and flooding in this region, leading to rapid sediment erosion from the surrounding mountains. This erosion results in thick layers of light gray sediment being deposited on the lake bottom. Rodbell, Seltzer, and their team analyzed these layers in long sediment cores taken from Lake Pallcacocha to calculate the frequency and intensity of past El Niño events. They found that, prior to 5,000 years ago, during the early to mid-Holocene, flood layers occurred on average every fifteen years, whereas during the late Holocene, flood layers occurred every two to ten years. This result suggests that El Niño events have been occurring more frequently in recent times.

VARIABILITY IN THE PACIFIC
DECADAL OSCILLATION

Researchers have also looked for clues to past climate changes in the strength of the Pacific Decadal Oscillation (PDO)—another pattern of ocean-atmosphere interaction that influences precipitation in the American West (as described in chapter 4). When the PDO is in a "negative," or cool, phase, sea surface temperatures are typically cooler than average; conversely, when the PDO is in a "positive," or warm, phase, sea surface temperatures are warmer than average. Researchers who use paleoclimate data to reconstruct past sea surface temperatures are able to see patterns of change in the phase of the PDO and even in the strength of the PDO. Such studies suggest, for example, that the PDO was generally in the negative (cool) phase during the middle part of the Holocene, about 7,000 to 4,000 years ago, so sea surface temperatures along the Pacific coast were cooler. Conditions during this period were wetter in the Pacific Northwest and drier in the Southwest and interior western North America (as described in chapter 7). The wetter conditions of the Neoglacial that followed (as described in chapter 8) correspond to a period when the PDO was in a positive (warm) phase.

The PDO returned to the negative (cool) phase again in the late Holocene for the four centuries from AD 900 to 1300. As we described in chapter 9, this was a period marked by extreme and prolonged droughts. The connection between the PDO phase and this megadrought period has been inferred from tree-ring records from Southern California and western Canada—which are at the extreme ends of the North American region affected by the PDO. Tree rings from these regions reveal that the Southwest was dry and the Northwest was wet at that time. A study using moisture-sensitive tree-ring chronologies from Southern California and northern Baja California has shown that the PDO was unusually weak during the Little Ice Age compared to that of the centuries immediately before and after it.

Although ocean conditions of the past seem to explain the broader patterns of precipitation in the West, other hypotheses have been put forward to explain some of the more frequently recurring climate cycles that we see in the past. The more intriguing of these hypotheses are ones that involve extraterrestrial causes of climate change.

EXTRATERRESTRIAL CAUSES OF CLIMATE
CHANGE: SUNSPOT CYCLES

Whereas the PDO and the ENSO are climate oscillations produced internally within the earth's ocean-atmosphere system, our planet's climate is also influenced by extraterrestrial events, including changes in the output of energy from the sun. The study of solar variations has a long history, extending back to the early seventeenth century when Galileo Galilei systematically studied the sun at dawn and dusk with his rudimentary telescope. He observed dark patches on the sun's surface, carefully recording their locations and numbers, and noticed that the number and positions of the dark patches shifted from year to year. Galileo and other astronomers recorded a ten- to twenty-fold decrease in these dark spots beginning in AD 1645. For the next seventy years, between AD 1645 and 1715, the sun's surface was almost devoid of sunspots.

In the late nineteenth century, the sunspot records were reexamined by astronomer Edward Walter Maunder. This period of low sunspot numbers from AD 1645 to 1715 is now known as the "Maunder Minimum." Maunder noted that this period coincided with the peak of the Little Ice Age, the cool interval between the fifteenth and nineteenth centuries, and he proposed a relationship between sunspots and climate.

Two other periods of low sunspot activity occurred during the past millennium: the "Wolf Minimum," peaking at AD 1300, and the "Sporer Minimum," peaking at AD 1500. These, and the Maunder Minimum, are spaced about 200 years apart, remarkably similar to the recurrence intervals of extreme climate events—floods and droughts—in the American West and elsewhere across the globe. In the West, these extreme events were recorded in the stratigraphic records of megafloods in the Santa Barbara Basin, changing lake levels of Mono Lake, and fluctuating salinity in the San Francisco Bay, to name just a few.

The pattern recurs beyond the western United States. For example, severe floods and droughts have been found to recur every two centuries in Central America. The Mayan empire of the Yucatan, which had flourished for thousands of years, was buffeted by periodic droughts and ultimately collapsed at the peak of its power nearly 1,000 years ago. The Mayans had long kept records of astronomical phenomena, and the droughts coincided with a pronounced 208-year cycle of increased solar intensity, a fact they may have noted with grave concern.

The Sunspot-Climate Connection

Although the connection between sunspots and climate is intriguing, the causal mechanism remains somewhat mysterious. The change in solar energy received by the earth between solar sunspot maxima and minima is a mere one-tenth of 1 percent. How could such a small change cause such relatively large climate shifts on the earth?

Sunspots appear when loops of magnetism form deep within the sun and rise to the surface, leading to a drop in temperature that appears as a dark spot when viewed from the earth. During periods with more sunspots, solar output is greater, with more solar energy being received at the top of the earth's atmosphere. Satellite observations have shown that the solar constant—the total amount of solar energy flux received by the earth—is not actually "constant." Rather, it varies on timescales of days to years, apparently related to the number of sunspots. Climatologists (and economists) have noted an eleven-year sunspot cycle that has been correlated with such climate-related phenomena as agricultural production in the United States. During periods of high sunspot activity, the value of the solar constant increases, triggering a chain reaction in the atmosphere and ocean that ultimately leads to changes in precipitation and temperature patterns in different regions of the earth.

Scientists are now exploring the details of changing solar output and associated effects on the earth's climate system. They use climate models to show the effects of small energy changes received in the upper atmosphere. These changes can be amplified through positive feedbacks in the planet's climate system. Climate modeler Gerald Meehl and his colleagues at the National Center for Atmospheric Research in Boulder, Colorado, have proposed two mechanisms for this amplification. In the "top-down" mechanism, increased sunspot activity leads to an increase in the amount of ultraviolet light hitting the stratosphere—enhancing the formation of ozone, absorbing more ultraviolet light, and thereby heating the stratosphere. The amount of heating in the stratosphere is not uniform across the globe but varies with latitude. The temperature differences lead to an increase in winds and tropical precipitation.

In the second, "bottom-up," mechanism, greater sunspot numbers lead to increased absorption of solar energy by the surface ocean in the subtropics and so increased evaporation of seawater. The trade winds become stronger, and the evaporated moisture carried by the trade winds increases precipitation in the western Pacific and subtropical zone. These stronger winds blowing from east to west across the Pacific increase the upwelling of cooler water in the eastern equatorial Pacific, lowering sea surface temperatures, similar to conditions during a La Niña event.

Although the mechanisms linking sunspots and climate are still being investigated, these climate models are important steps in understanding why precipitation patterns in the American West and elsewhere appear to be closely tied to sunspot cycles.

Radiocarbon and Past Solar Activity

In an attempt to understand past changes in solar output in deeper geological time, Earth scientists have devised ingenious proxy methods for solar activity. One of these methods is the amount of radiocarbon produced in the earth's atmosphere, which is closely related to the output of energetic particles from the sun (the solar wind). Periods of low solar activity correspond to high radiocarbon production in the earth's atmosphere and vice versa. Radiocarbon (carbon-14) combines with oxygen to form carbon dioxide and is taken up by plants and trees during photosynthesis. Years with higher levels of radiocarbon production will thereby be recorded in the growth rings of the trees. Geochemists measure the minute amounts of carbon-14 in tree rings and compare this radiocarbon age to the actual (calendar year) age of the tree rings, obtained by counting the annual rings.

The tree rings used in these studies are those of the long-lived bristlecone pines in the White Mountains of eastern California, with living trees that are 4,000 years old and fossilized trees extending the record back to 11,000 years before the present. These studies have shown that, in addition to a 200-year cycle in solar activity, there are solar cycles with periods of 55, 90, 400, 1,000, and 2,500 years. One key finding is that these periodicities are seen in a number of climate records in the West, including extreme floods and droughts, suggesting that solar activity somehow affects when and where precipitation falls by influencing atmospheric circulation.

Another cycle that climate scientists have detected has a 1,500-year periodicity. For decades, geologists have noted that glaciers throughout the Holocene advanced and retreated every 1,500 years or so. More recently, this 1,500-year cycle has also been documented in oceanic sediment cores in the form of discrete layers of debris transported to the North Atlantic by icebergs (known as "ice-rafted debris"). In the eastern Pacific, changes in sea surface temperatures from the Santa Barbara Basin similarly exhibit a 1,500-year cycle. Thus, this cycle appears to correlate with solar activity, but details of how it influences climate are still being investigated.

TERRESTRIAL CAUSES OF CLIMATE CHANGE: VOLCANIC ERUPTIONS

Volcanic eruptions—violent natural events in their own right—have been linked with past climate variations. Although eruptions occur relatively frequently, they are not cyclical. Over timescales of millions of years, changes in the number of volcanic events on land and undersea (where volcanism occurs at mid-ocean ridges and during the formation of oceanic plateaus) eventually alter the flux of carbon dioxide in the atmosphere, modifying climate. For instance, the Mesozoic Era, when dinosaurs roamed the earth, had a much balmier climate primarily because of increased undersea volcanism associated with faster seafloor spreading and oceanic plateau formation. This volcanism increased the amount of carbon dioxide from the interior of the earth to the atmosphere, resulting in atmospheric levels higher than 2,000 parts per million and high global temperatures.

Over shorter timescales of months to years, volcanic eruptions have had the opposite effect on temperatures. Following an eruption, particles of dust and ash are ejected into the atmosphere, where they are suspended, blocking sunlight and reducing solar radiation for a year or more. Volcanoes also eject sulfur gases, which combine with water vapor in the stratosphere to form tiny droplets of sulfuric acid. These droplets absorb solar radiation and scatter it back into space, causing global temperatures to fall.

Volcanologists have compared the climate effects of recent eruptions and observed that the greatest cooling is caused by sulfur-rich eruptions. The 1982 eruption of El Chichon in Mexico, for example, emitted a smaller volume of ash than the 1980 eruption of Mount St. Helens in Washington but forty times the volume of sulfur gases, and El Chichon resulted in five times the amount of atmospheric cooling. Although these effects are relatively short-term, they can be significant enough to have bigger climatic impacts. Two major volcanic eruptions are thought to have contributed to global cooling during the Little Ice Age. In Iceland, the Laki eruption in 1783 was the largest basaltic eruption in recorded history, with 3.4 cubic miles of lava erupting over a period of eight months. The eruption caused extremely severe winters in Europe and North America. At the time, Benjamin Franklin observed a constant fog over much of North America and Europe, and he wondered whether the unusually low winter temperatures experienced that year in the eastern United States, where temperatures were 8°F lower than they had been the previous 225 years, were related to the Laki eruption.

Three decades later, an even larger eruption (the largest of the past 10,000 years) occurred when Mt. Tambora erupted in Indonesia in 1815. Some 36.5 cubic miles of ash were sent up into the atmosphere, resulting in what became known as the "year without a summer," featuring unusually low temperatures, frost, and snow even during summer months in Europe and New England. The amount of ash spewed into the air by Mt. Tambora was about a hundred times that of Mount St. Helens in 1980.

The relatively rapid climate fluctuations of the past 11,000 years are super-imposed on more gradual changes in the orbit of the earth around the sun. The earth's climate has fluctuated in and out of ice ages over the past two and a half million years, as described in chapters 5 and 6. What caused these recurring ice ages, how were they discovered, and how have they affected climate during the Holocene?

The theory of the nature and causes of these longer-term variations in climate has evolved over the past two centuries with the help of a distinguished group of astronomers, geologists, mathematicians, climatologists, and physicists. One of the earliest and most unlikely, and thus remarkable, contributors to our understanding of these cycles was James Croll. He grew up in Scotland in the early nineteenth century and was forced to drop out of school at the age of thirteen to work on the family farm. After his long and arduous days in the fields, Croll returned home to study physical science late into the night. He had a passion for understanding the forces governing the natural world. His family could not afford a formal university education, so, when Croll reached his late teens, he was forced to embark on a series of jobs in which he had little interest or aptitude, including millwright, carpenter, salesman, shopkeeper, hotel keeper, and life insurance salesman. Only in his spare time could he study his true passion—earth science.

A major turning point for Croll came in middle age, when he accepted a job as a janitor at the Andersonian College and Museum during the 1860s. His salary was meager, but he had access to an extensive scientific library where he spent evenings studying physics and geology. He was drawn to a hotly debated topic of the day: the features and causes of past ice ages. He was particularly interested in the hypothesis that past ice ages were related to changes in the earth's orbit around the sun.

The shape of the orbit itself changes on long timescales, pulsing from near-circular to slightly more elliptical. The gradual distortion of the earth's orbit is caused by the gravitational pull of other nearby planets. When the orbit is elliptical, the earth receives more sunlight than at other times because of the slight change in its distance from the sun. Today, the earth's orbit is very close to circular, and there is only a small difference in the amount of radiation received during its yearlong revolution.

Croll calculated that, for each revolution of the earth around the sun, the amount of solar radiation received on the earth's surface is slightly different than it was the year before. The difference is infinitesimal but cumulative. He speculated that these small changes in the earth's orbit could alter the amount of solar radiation reaching the earth's surface just enough to influence the earth's climate. His laborious calculations of the earth's orbit over the past three million years showed cyclical variations between more circular and more elliptical shapes, with periods lasting 100,000 years.

Another aspect of the earth-sun relationship also interested Croll. He hypothesized that maximum ice growth, leading to an ice age, would occur when the orientation of the earth's axis of rotation was in a certain position. The earth's axis of rotation also changes cyclically; it "wobbles" like a spinning top. Today, the earth's axis at the North Pole points toward the North Star. Over time, however, the orientation of the earth's axis slowly changes, tracing a circle in the sky that takes 26,000 years to complete. Croll reasoned that the most likely time for the growth of a large ice sheet in the northern hemisphere (where most of the global landmass is located) is when the earth's orbit is most elongated and when the axis of rotation is pointing away from the sun during the winter.

Although Croll's calculations showed that the earth's orbital changes caused only slight variations in the amount of solar energy reaching its surface, he surmised that ice sheets, once they begin to grow, reflect sunlight back to space, causing even more cooling and resulting in more ice—a so-called "positive feedback" loop. Croll can be credited with being the first to recognize this important process. He also proposed that growing ice sheets near the poles would increase the temperature differences between the equator and poles, leading to more vigorous winds and surface currents, ultimately bringing more moisture toward the poles. These processes would enhance ice sheet growth and are still recognized as important today.

Despite his humble beginnings, his lack of formal education, and his need to work long days in a variety of jobs ill-suited to him, Croll became a prominent scientist, publishing a book entitled *Climate and Time in Their Geological Relations: A Theory of Secular Changes of the Earth's Climate*. He finally landed a position at the Scottish Geological Survey, and ultimately his impressive achievements were honored by his election to the Royal Society of London.

By the late nineteenth century, however, Croll's ice age theory was questioned and finally rejected. His proposed date for the end of the last ice age, some 80,000 years ago, was shown to be off by tens of thousands of years. A much later date, 10,000 years, was estimated by geologists who studied the erosion of young glacial deposits by the Niagara River. This disparity led the geologists at the time to reject Croll's entire theory, and, sadly, he died an unappreciated scientist.

MILANKOVITCH CYCLES

Fortunately, Croll's ice age theory was revived two decades later by a young Serbian mathematician, Mulatin Milankovitch, who had read Croll's book and was so inspired that he made it his life's goal to understand the earth's past climate. He began with calculations of the amount of solar radiation received by the earth's surface at different latitudes and seasons over the past 130,000 years, based on the orbital cycles. Milankovitch calculated that variations in the tilt of the earth's rotational axis seemed to produce the largest changes in solar radiation (and temperature) on the earth's surface. The tilt of the spin axis with respect to the earth-sun plane, currently 23.5°, fluctuates between 22° and 24.5°. The importance of this becomes more evident when we consider that the earth's tilt controls our seasons: when the Northern Hemisphere tilts away from the sun's rays, it experiences winter, and when the Northern Hemisphere tilts toward the sun, it experiences summer. Each day, the earth completes a rotation around this axis that is slightly different than the day before. The difference again is tiny, but it forms a tick in a long cycle that takes 41,000 years to complete, during which our seasons become more or less pronounced.

Contrary to Croll's findings, Milankovitch calculated that the most likely time for the growth of large ice sheets is not during the Northern Hemisphere winter but during its summer. When summer temperatures in the mid- to high latitudes of the Northern Hemisphere are cooler, the snow that has accumulated over the winter persists during the summer, rather than melting. Year by year, snow continues to build up, finally turning to ice under the accumulated weight and pressure, eventually growing into a large ice sheet.

EVIDENCE FROM THE DEEP SEA

The final link in the great ice age puzzle was discovered in the 1950s, when geologists began drilling cores in the sediments of the deep seabed. These sediments include tiny carbonate microfossils (the foraminifera) that continuously rain down through the water column. The microscopic fossils allowed geologists to derive the pattern of the ice ages for the past two million years.

Pioneering research measuring the oxygen isotopes in tiny fossil shells was conducted by Cesare Emiliani in the laboratory of his mentor, Harold Urey, at the University of Chicago. Emiliani developed the use of oxygen isotopes in paleoclimate research—a technique whose value, as we have already seen, continues to this day. It is based on the more rapid evaporation of the lighter isotope of oxygen (oxygen-16) from the ocean than the heavy isotope (oxygen-18). This evaporated water vapor is carried to higher latitudes and continents, some precipitating as snow on land, building great ice sheets during glacial periods. As the oxygen-16-rich water is removed from the oceans and stored as ice on land, sea level is lowered, and the ocean water becomes more enriched in the heavy isotope of oxygen (oxygen-18). When marine organisms (such as the microscopic foraminifera) build their shells of calcium carbonate, they incorporate this oxygen, recording changes in sea level and glacier growth on land.

Emiliani published a landmark paper in 1957 that described oscillations every 100,000 years between two opposing climate states: cold glacial periods, lasting up to 90,000 years, and warmer interglacial periods, lasting only about 10,000 years. The temperature difference between the two states is approximately 6°C (11°F) or more. The shift from the warmer to the cooler state into an ice age is relatively slow, as ice builds up on the Northern Hemisphere continents over the course of tens of thousands of years. In contrast, the shift from the cooler to the warmer state (from glacial to interglacial) takes only thousands of years. This pattern of climate change over the past million years has been described as "sawtooth" because of its sharp up-and-down shape when temperature or sea level is plotted against time.

EVIDENCE FROM ICE CORES

Ice sheets on Greenland and Antarctica contain in their layers of ice detailed information about past global temperatures, atmospheric composition, windiness, and more. Over the past two decades, cores extracted from these enormous ice sheets have been used to reconstruct past climate. In one project, researchers extracted atmospheric gases from bubbles trapped in the ice to measure past levels of carbon dioxide. Their results show that carbon dioxide levels in the atmosphere varied with temperature during the ice ages over the past two million years. Carbon dioxide dropped to concentrations as low as 180 parts per million (ppm) during the peak of the glacials and rose to about 280 ppm during the interglacials.

Carbon dioxide causes a climate feedback by amplifying the small changes in solar radiation received by changes in the earth's orbit. During glacial periods, carbon dioxide is absorbed by the oceans, decreasing its concentration in the atmosphere, leading to further cooling, in a positive feedback loop. During the transitions from glacials to interglacials, carbon dioxide is released back to the atmosphere, further warming the earth, in yet another positive feedback loop. Carbon dioxide levels reached their "interglacial" value of 280 ppm about 6,000 years ago but have climbed to the unprecedented level (over the past two million years) of 390 ppm over the past century. Many climate scientists warn that a continued rise in atmospheric carbon dioxide will warm the earth's climate for the foreseeable future.

The ice cores also show that, during glacial periods, the climate changed quite abruptly—warming rapidly and then cooling—every few thousand years. Although we have not experienced such drastic climate swings during the Holocene, climate scientists are actively studying these past events in order to understand what caused them and to determine if they might recur in a future warmer world.

THE LAST BIG THAW

At the peak of the last glacial period, some 20,000 years ago, ice sheets reached a maximum size (and sea level its lowest level) during a period called the Last Glacial Maximum. Solar radiation in the Northern Hemisphere summer dropped to a minimum at that time, and the global climate was as cold as it had been in 100,000 years. Gigantic ice sheets blanketed North America, Europe, Asia, and Scandinavia. These ice sheets were between one and two miles thick; their weight was so great that they depressed portions of the continents by as much as 1,200 feet. The water locked up in these enormous ice sheets originated from the oceans, causing global sea levels to drop by about 360 feet.

Gradual shifts in aspects of the earth's orbit explain the thawing and melting of the ice ages beginning 20,000 years ago as well as the broad climate patterns that have occurred since then. Summer solar radiation in the Northern Hemisphere began increasing from a minimum 20,000 years ago to peak values about 10,000 years ago, when the earth's tilt reached its maximum of 24.5°. However, these astronomical conditions were not enough to account for the 18°F difference between glacial temperatures and Holocene interglacial temperatures.

Climate feedbacks are required to accentuate these nudges in global conditions. As the glaciers retreated, the white cover of snow and ice, which reflects most of the sunlight, was removed from the continents, replaced with a darker earth surface. That darker surface absorbed more of the sunlight, transferring the heat to the surrounding area rather than reflecting it back into space. During the early Holocene, once the southern edges of the ice sheets began to retreat, the earth's reflectivity (or albedo) decreased, with less solar radiation reflected back into space and more of it retained by the earth's surface, where it warmed the land and melted more ice. Many such feedback loops across the globe have worked together in leading the earth into the ice ages and back out again.

ENORMOUS MELTWATER LAKES AND MEGAFLOODS

The melting of enormous ice sheets led to the formation of vast glacial lakes. Rather than draining directly to the sea, the meltwater from the ice sheets sometimes became trapped behind ice dams and formed gargantuan lakes. As these lakes grew, they periodically burst through their dams, forming some of the largest floods in the earth's history.

Evidence for these enormous glacial floods was first discovered by a young field geologist named J. Harlan Bretz in the early twentieth century. As described in a book by Doug Macdougall, Bretz spent his summers during the 1920s and 1930s documenting and mapping the enigmatic geology of eastern Washington and Idaho. He found gigantic boulders sitting on the landscape 650 feet higher than the present-day Columbia River; intertwining channels and canyons cut into hard basalt bedrock to depths of 1,000 feet; enormous, extinct waterfalls; and giant gravel bars and potholes. These seemingly unrelated features began to tell an almost unbelievable story: this region of ripped-up earth in eastern Washington, the so-called "Channeled Scablands" (see figure 18 in chapter 7), appeared to be the result of a megaflood of unimaginable proportions—possibly the largest ever to sweep across the face of the earth.

Bretz hypothesized that this megaflood originated in northwestern Montana, at the southern boundary of the ice sheet two miles thick—now

called the Laurentide ice sheet—that blanketed Canada and the north-
ern United States at the end of the last ice age. As the earth warmed, the
Laurentide ice sheet melted. The waters that had been locked in the ice were
released to make their way back to their place of origin—the ocean. In North
America, these waters took several routes: east through the St. Lawrence
Seaway to the North Atlantic; south through the Mississippi River to the
Gulf of Mexico; and west through the Columbia River to the Pacific Ocean.

One of the largest of the glacial lakes was named Missoula, and it covered
much of northern Montana. Even as Bretz was mapping giant flood features,
a glacial geologist named Joseph Thomas Pardee was studying the sediments
deposited at this lake's bottom. As Macdougall's book also describes, Pardee's
work revealed that Lake Missoula was once the size of Lake Michigan
but even deeper: ancient lake shorelines mapped in the hills surrounding
Lake Missoula revealed that its depth was, in places, 2,000 feet. Linear hills
45 feet high and separated by over 400 feet were enormous ripple marks
that formed just west of Lake Missoula under the huge volume of floodwater
draining the lake.

Putting together all the evidence, Bretz hypothesized that an ice dam once
held back the lake, and, when that dam melted, the water trapped behind it
burst forth in a catastrophic flood. The sound of the bursting dam would
have been deafening as it rumbled for hundreds of miles ahead of a wall of
water 500 feet high pouring across the Columbia Plateau toward the Pacific
Ocean. It would have swept up everything in its path, a brown and churning wave of water, soil, vegetation, and floating icebergs still embedded with
boulders. The discharge rate of this massive flood was likely over 200 times
that of the largest historical floods on the Mississippi River.

This terrifying flood was probably the largest but not the only one that
occurred in the Northwest as the earth thawed from the last ice age. Evidence
for multiple flood events comes from the layers of sediment accumulated
in the Scablands. The scoured debris and sediments in these floodwaters
were eventually carried out to the Pacific Ocean to a final resting place over
600 miles from the Oregon coast, where they were recently cored and
recognized as further evidence for the megafloods.

As we have seen, a number of factors influenced climate over different
timescales. The earth's climate is affected by many powerful processes,
including changes in its orbit, changes in solar output, volcanic eruptions,

and oceanic changes. These factors act simultaneously and involve feedback loops, producing changes in regional and global climate. For regions like the American West, the net result over the past 20,000 years has been a climate that fluctuates broadly between warmer and colder, wetter and drier, and is punctuated by more extreme floods and droughts that recur on a regular basis.

The average climate in the West has been relatively benign over the past century and a half, encouraging human settlement and expansion. A certain faith in the reliability of water needed for the quality of life we enjoy has also fostered a tremendous population growth. In the following chapter, we present an overview of how humans during the past 150 years have adapted to the climate in the West and the ways in which these adaptations, including extensive water development projects, have affected ecosystems in the West, particularly those in the aquatic environment.

PART THREE

A Growing Water Crisis

We are in a raft, gliding down a river, toward a waterfall. We have a map but are uncertain of our location and hence are unsure of the distance to the waterfall. Some of us are getting nervous and wish to land immediately; others insist we can continue safely for several more hours. A few are enjoying the ride so much that they deny there is any immediate danger although the map clearly shows a waterfall. . . . How do we avoid a disaster?

S. GEORGE PHILANDER, *Is the Temperature Rising?*

TWELVE

The Hydraulic Era

SALMON AND DAMS

> Wild salmon are survivors—living through volcanic eruptions,
> ice ages, mountain building, fires, floods, and droughts. I feel
> certain they will persist if we can control our behavior and give
> back what we've taken from them—rivers that retain some of
> their healthy ecological function.
>
> JIM LICHATOWICH, *Salmon without Rivers*

THE TWENTIETH CENTURY HAS BEEN called the "hydraulic era" in the American West. In fact, it took less than a century for water engineers, obsessed with providing enough water for an ever-growing population, to harness nearly every drop of naturally flowing water in California. Once sporadic and untamed, the water resources not only of the state but also throughout the American West were rapidly controlled, and the natural hydrology was re-engineered through massive public works including dams and aqueducts. Today, residents of the western states are linked to each other by an extensive web of aqueducts and pipes that bring them water every day. Dams are the modern solution to climate variability and unpredictability—trapping the water that falls during the winter months or melts in the spring, then storing and transporting that water to where it is needed during the drier summer and fall months. In addition to providing a steady flow of water to residents and agriculture, these water works also provide hydroelectric power and protection from flooding.

It has been argued that water engineering during the hydraulic era was a response to what water planners saw as a need to keep pace with a steadily increasing population and expanding agriculture across the American West. But others argue that the massive engineering of the region's water resources was precisely what led to the rapid population growth, allowing it to grow to a size far larger than the region could support naturally.

Regardless of cause and effect with respect to population, the fact remains that the monumental changes to the region's natural rivers, lakes, and wetlands have wreaked havoc on the natural aquatic ecosystems that these waters harbored. The plants and animals that depend on streams, lakes, and wetlands declined rapidly over the twentieth century due to large-scale land and water development in the West. Hundreds of species that are federally listed as threatened or endangered depend on rivers and streams. Over three-quarters of the native freshwater fish in the West are either extinct already or listed as endangered.

Human existence in the West relies not just on water resources but also on these other natural resources—the plants and animals—that the water sustains. The West is at a crossroads today as the human population in the region continues to increase, controlling ever more of the limited freshwater at the expense of aquatic ecosystems. Massive and profound as these public works are, they have been part of life in the West for so long that people are almost unaware of them. Few would voluntarily go back to the pre-hydraulic era, when water was unreliable from year to year and much of the agricultural industry would not have been possible. However, water development has come with a steep price to the environment—particularly to aquatic ecosystems—and the once-mighty salmon now serves as its poster child. In this chapter, we recount several of the events of the twentieth century that led to massive reconfigurations of water in California and discuss the impacts of water development on the state's environment. We use the chinook, or king, salmon (*Oncorhynchus tschawytscha*)—a keystone species whose decimation over the past century has been closely linked to water development—to illustrate the effects of water development on aquatic ecosystems in general.

THE HYDRAULIC ERA BEGINS

During the westward push by pioneers in the early twentieth century, residents across much of the West experienced the region's unpredictable climate, with its periodic droughts and floods, and set about taming the rivers. They dug canals to water their crops, in some instances using ancient Native American irrigation canals that had been preserved for centuries. As the population grew, these attempts to control the water expanded, ultimately leading to a total transformation of the natural hydrology of the region:

FIGURE 30. Map of California showing aqueducts that bring water to Southern California, including the Los Angeles Aqueduct from the Owens Valley, the California Aqueduct from Northern California, and the Colorado River Aqueduct from the California-Arizona border. (Map redrawn by B. Lynn Ingram.)

water was stored and transported to whereever they wanted it, whenever they needed it.

In California, when people flocked to the abundant sunshine and mild climate of the southern part of the state, the lack of adequate water soon became apparent, particularly during periods of drought. The solution was to build the world's largest and longest aqueduct for the time, diverting river water from the Owens Valley to Los Angeles, 230 miles south (see figure 30). The Los Angeles Aqueduct carried pristine Sierra Nevada snowmelt that fed the Owens River, water praised by John Muir, who wrote in 1901: "It is not only delightfully cool and bright, but brisk, sparkling, exhilarating and so positively delicious to the taste that a party of friends and I led to it twenty five years ago still praise it, and refer to it as 'that wonderful champagne water;' though, comparatively, the finest wine is a coarse and vulgar drink" (p. 185).

As it turned out, only a small amount of this pristine water was actually used by Los Angeles residents for drinking. Most of that "wonderful champagne water" was used to irrigate orchards, fill swimming pools, and water lawns, gardens, and palm trees planted by the city in its effort to transform this semiarid Southern Californian desert into an empire.

The Owens Valley water project set the tone for the water battles that followed in the twentieth century. Disputes erupted between the Owens Valley farmers, who were eventually ruined as their water disappeared, and the rapidly expanding city of Los Angeles. Major environmental impacts resulted, including the disappearance of Owens Lake, which lay at the southern end of the Owens River, in 1924. For millennia, this ancient lake had been large (twelve miles by eight miles) and had served as an important nesting and feeding stopover point for millions of waterfowl along the Pacific Flyway. When the lake completely dried up in the 1920s, 108 square miles of dry lakebed, including a mixture of evaporite minerals (those that precipitate from the dissolved salts in the evaporating water), sand, and clay, were exposed to erosion. Winds blowing in the region generated great alkali dust storms, leading to poor air quality. The Owens Valley was rapidly converted from the lush agricultural region the pioneers had created back into a semi-arid desert with grasses and shrubs—albeit one without a river running through it.

Two decades later, Southern California again looked far afield for more water to meet its growing population: this time the Colorado River was to be its new water source. The 1931 propaganda film "Thirst" aimed to convince voters to approve bonds to build another aqueduct that would ultimately bring water to Los Angeles from the Colorado River. In this movie, the chairman of the Metropolitan Water District of Southern California narrates:

> All of Southern California was at one time a desert waste. We have reclaimed this desert and now we have in its place this growing empire. But the desert is ever around us, waiting and eager to take back what was once its own. And it will take it back, unless we bring in more water. Unless we take immediate steps to bring in water from an outside source, the people of Southern California will be up against a serious water shortage. But we're fortunate at having within our reach a water source capable of supplying our needs. This source is the Colorado River. We must not think that this is something for the future; it is of the most immediate and pressing necessity. If we are to survive and to grow, we must have the water that will enable us to maintain our mastery over the desert.

The film goes on to describe the region's rapidly declining water table, the result of more groundwater being pumped out of the ground than could be replenished by rain and runoff. The region was pumping an estimated 170 million gallons of groundwater per day. It was said, however, that the Colorado River could provide one billion gallons of water per day—over six times the amount they actually needed at the time. It is not surprising, then, that during California's prolonged and deep Dust Bowl drought voters approved a bond to fund the 242-mile Colorado River Aqueduct, which was completed in 1941.

In Northern California, the city of San Francisco began looking eastward to the Sierra Nevada for a more reliable source of water in the early twentieth century. The city set its sights on the Tuolumne River and proposed to dam the Hetch Hetchy Valley in Yosemite National Park. A battle between the Sierra Club (led by John Muir) and the city of San Francisco (with Gifford Pinchot, the first chief of the U.S. Forest Service) lasted between 1902 and 1913. In 1913, the city was authorized to build a dam on the Tuolumne River, transforming the Hetch Hetchy Valley into a large reservoir for San Francisco and other Bay Area cities—a project that would not be fully completed until 1934.

Two decades later, the California Aqueduct was built to bring water from the wetter northern half of California to the San Joaquin Valley and Los Angeles. This project was part of the larger "Central Valley Project" funded by the Roosevelt administration during the Great Depression. These twentieth-century water projects led to a population explosion in California, from 500,000 in 1920 to over 37 million in 2012.

In the Southwest, much of its population growth would not have been possible if not for water development, particularly of the Colorado River. The Colorado drains rain and meltwater from states in its upper drainage basin—Wyoming, Colorado, Utah, and New Mexico—and brings this water into the deserts of the lower basin, including Nevada, Arizona, and California (see figure 31). In the past, it was a capricious river: a sleepy stream one day but (following a winter storm or a summer monsoon downpour) a raging torrent the next.

The taming of the Colorado River began with the construction of Hoover Dam in the 1930s, followed by numerous other dam projects on the Colorado in the following decades. These water projects have reduced devastating flash floods but have also decimated the native fish species that were adapted to the natural conditions on the river. The Colorado now provides water and

FIGURE 31. Map of the southwestern United States showing the Colorado River and its drainage basin (shaded region). (Map redrawn by B. Lynn Ingram.)

hydropower for the largest cities in the West: Phoenix, Las Vegas, Denver, Los Angeles, and San Diego.

Every California river has at least one dam, and many have more than one, with 1,500 dams in this state alone. The reservoirs behind these dams hold almost 42 million acre-feet of water, which is about 60 percent of the state's total yearly river runoff. Throughout the twentieth century, enormous water projects were built by federal and state governments in California, including the Central Valley Project, the State Water Project, San Francisco's Hetch Hetchy Project, the Sacramento–San Joaquin Flood Control Project, the Colorado River Aqueduct, and Los Angeles's Owens Valley Project. Water is captured and stored in reservoirs, then transported from the snow-capped Sierra Nevada range to the drier Central Valley farmlands and Southern Californian cities, where about 60 percent of the population lives. This "water on demand," in addition to extremely fertile soils, has transformed California's Central Valley into the "bread basket of the world"—growing 55 percent of America's produce.

The twentieth-century era of dam construction eventually came to an end in the 1970s as the environmental impacts of water development could no longer be ignored. The extensive system of dams and aqueducts in the American West had been built when the interconnections between waterways and the ecosystems they support were poorly understood and not yet in the popular press. Since then, research and restoration has begun in an attempt to save aquatic ecosystems that are on the verge of collapse. Today, the flows left in rivers to restore aquatic ecosystems—dubbed "environmental flows"—are a hotly debated topic between farmers, fishermen, environmentalists, and scientists.

Healthy watersheds—and the ecosystems they support—go largely unappreciated because they lack a market value. Nevertheless, rivers provide food, materials, and recreation, and freshwater ecosystems regulate environmental conditions that benefit humans. For instance, wetlands improve water quality by filtering out pollutants and impurities, and they provide an important buffer during periods of flooding.

ENVIRONMENTAL COSTS OF THE HYDRAULIC ERA

Declining diversity in freshwater habitats in the western United States is largely a result of the combined effects of natural hydrologic patterns, pollution, and nonnative species invasions. In California, fish provide the best indicators of aquatic deterioration. Of the 129 native fish species in the state, 40 percent are in decline, 37 percent are eligible for listing as threatened or endangered, and 5 percent are extinct. Most of these species (60 percent) are found exclusively in California, so their decline is thought to be due to factors within the state, directly related to water and land development.

The Salmon Story

The plight of the once prolific salmon on the Pacific coast illustrates the impact of water development on aquatic life. Salmon have important linkages with many ecosystems from the ocean to streams and are therefore considered a keystone species. The chinook salmon have evolved along with western watersheds over thousands to millions of years and have been a mainstay of coastal Native American diets for at least five millennia. Although natural population sizes of salmon have fluctuated over this time period, at no point in the past have their numbers dropped as low as they have over the past century.

The natural range of chinook salmon encompasses the northern Pacific Ocean from California to the Gulf of Alaska and includes entire watersheds of the western states: from the rain forests of the Cascades in the Northwest to the Sierra Nevada range and deserts of California and as far inland as Rocky Mountain streams. Salmon by the millions once migrated up rivers along the Pacific coast, transforming streams into sparkling "silver pavement."

In his book *King of Fish: The Thousand-Year Run of Salmon,* geomorphologist David Montgomery at the University of Washington describes how salmon migration patterns and speciation co-evolved with the geomorphology of western North America over the past twenty million years. During this time, the plateau between the Rocky Mountains and the Pacific fragmented into the fault-blocked basin and range topography in Nevada, while the mountain ranges along the Pacific coast, from California to Alaska, were uplifted. These topographic changes inland and along the coast formed many disconnected watersheds with distinct stream conditions. The isolation of these streams eventually led to the evolution of different salmon species.

For many millennia, the streams and rivers of these coastal watersheds have provided breeding grounds for the salmon. Hatching in the cold, rapidly running mountain streams, the young salmon fry develop into juvenile salmon and then make their way downstream and eventually out to sea. There, they spend their adult years in the open waters of the Pacific before beginning their arduous and improbable return-migration to their ancestral streams, where they lay their eggs in the cold water and gravel beds before they die. How they accomplish this remains to some extent a mystery, but these faithful fish are emblematic of the interconnectedness of conditions in the oceans, in the climate, and on land.

The Climate Connection

Salmon populations have fluctuated in response to climate change over the northern Pacific Ocean operating on a variety of timescales. When sea surface temperatures are unusually cold off the coast of the Pacific Northwest, the upwelling of nutrient-rich waters enhances the growth of phytoplankton and krill, allowing salmon populations to do well. During periods when reduced upwelling leads to higher surface water temperatures, krill and phytoplankton decrease, leaving less food for salmon and causing a decline in their populations.

Nate Mantua, a climatologist at the University of Washington in Seattle, has shown, with his colleagues, that salmon population size changed in Alaska and the Pacific Northwest in response to ocean conditions over the past century. They demonstrated that North Pacific sea surface temperatures were generally warmer from 1925 to 1947 and again from 1977 to 2000; between 1947 and 1977 and again after about 2000, sea surface temperatures were generally cooler. These shifts in ocean conditions are part of the Pacific Decadal Oscillation discussed in previous chapters. Here, we note simply that this climatic shift, or oscillation, has had direct effects on salmon productivity throughout this vast region.

Salmon populations have also responded to broader climatic shifts on longer timescales—like those described in chapters 9 and 10. For instance, a study looking at the chemistry in sediments cored from a Kodiak Island lake in Alaska has documented shifts in salmon abundance over the past two thousand years. During the three to five years that adult salmon spend in the ocean, they assimilate an isotope of nitrogen (nitrogen-15) in their tissues that is unique to the marine environment. After the salmon migrate back into the watershed (in this case, they migrate up the rivers into a lake), the nitrogen-15 from their tissues ends up in the lake sediments after the fish die and decompose. Analyses of nitrogen-15 in the lake sediments collected in layers provide a measure of the amount of salmon entering the lake from the ocean over time. The lake sediment cores reveal a pattern of fluctuating populations, with a low abundance of Alaskan salmon prior to AD 800, then increasing from AD 800 to 1200, and decreasing again after AD 1200. This pattern correlates with climate changes over the American West. During the Medieval Climate Anomaly (discussed in chapters 9 and 11), ocean surface temperatures off the coast were more often cooler than usual; these data correlate with a period of greater abundance of salmon as seen in the lake sediments. Subsequently, the isotope data in the lake sediments indicate that salmon populations dropped during the Little Ice Age, a time when ocean temperatures were often unusually warm, with frequent and intense El Niño events (see chapters 10 and 11).

Despite the natural fluctuations in salmon abundance associated with changing climatic conditions, salmon have been a mainstay for Native American populations along the Pacific coast for at least 5,000 years. Many tribes today still consider salmon critical for their survival as a culture and a people. The relationship between these tribes and salmon along the Pacific coast evolved as salmon habitats and annual runs became more stable and abundant during the mid-Holocene, when sea levels stabilized following the last ice age (as described in chapter 7). The native people understood the salmon's natural migration patterns and managed for a sustainable harvest, using barbed spears, harpoons, dip nets, and seine nets. Prior to European contact, Native Americans in California each year caught and consumed an estimated 15 million pounds of salmon. Farther north, on the Columbia River, a population of 50,000 Native Americans harvested between 20 and 40 million pounds of salmon annually—representing about 5–20 percent of

the annual salmon runs, depending on yearly variability in salmon populations. These harvests did not seriously affect the salmon over the thousands
of years that humans have lived along the Pacific Coast. Yet now, over the
past century, the salmon along the West Coast have almost disappeared.

Causes of Salmon Declines

Today, salmon are at the center of some of the most contentious environmental battles in the West, all swirling around water and land development.
The steady decline of the salmon along the West Coast began in the mid–
nineteenth century, when pioneers settled the region and began altering the
landscape. In California, hydraulic gold mining was the first major assault
on migrating salmon. Streams were choked with sediment that was washed
out of the Sierra Nevada foothills by enormous hoses. Moreover, commercial
fishing of salmon began in rivers in the 1860s and in the ocean in the 1890s.
But it was the twentieth-century era of dam building, the hydraulic era, that
had the greatest impact on salmon populations, reducing them to a small
fraction of their former levels. Other factors affecting the salmon—such
as logging and cattle grazing in the watersheds, pollution from abandoned
mines, gravel mining in streams, overfishing in the ocean, and periods of
natural climate variability with unusually warm ocean conditions—have also
contributed to their plight. But all of these factors combined are thought
to have played only a minor role compared with the severe impacts of dam
building on West Coast rivers.

Dams physically block salmon migration, preventing the fish from reaching their spawning habitat, and this is thought to be the primary cause of
salmon decline. In California, juvenile salmon that manage to reach the
Sacramento–San Joaquin Delta and the San Francisco Bay often fail to reach
the ocean because many are entrained in the huge battery of pumps in the
southern delta that transport water to the California Aqueduct and down
to the San Joaquin Valley and Southern California (see figure 30). The small
juveniles have evolved to follow the currents, and these pumps have reversed
the currents in the delta to flow south, away from the estuary and the Pacific
Ocean. Millions of salmon fingerlings are lost each year, sucked into the
pumps and killed, decimating future salmon populations. In addition, the
delta pumps draw a greater proportion of river water during droughts, when
farmers have a greater demand for water, leading to even higher fish mortality
during an already stressful time.

Although salmon populations have fluctuated over time as part of natural cycles, their numbers have reached an all-time low over the past several decades. Fisheries biologist Jim Lichatowich writes in his book *Salmon without Rivers:*

> The salmon are among the oldest natives of the Pacific Northwest, and over millions of years they learned to inhabit and use nearly all the region's freshwater, estuarine, and marine habitats. Chinook salmon, for example, thrive in streams flowing through rain forests as well as through deserts; they spawn in tributaries just a few miles from the sea and in Rocky Mountain streams 900 miles from saltwater. From a mountaintop where an eagle carries a salmon carcass to feed its young, out to the distant oceanic waters of the California current and the Alaskan Gyre, the salmon have penetrated the Northwest to an extent unmatched by any other animal. They are like silver threads woven deep into the fabric of the Northwest ecosystem. The decline of salmon to the brink of extinction is a clear sign of serious problems. The beautiful ecological tapestry that northwesterners call home is unraveling; its silver threads are frayed and broken. (p. 6)

In many ways, the salmon in Pacific coastal rivers reflect the health of the entire watershed: their decline is a warning sign that these ecosystems are suffering. The chinook salmon experienced unprecedented declines from 2006 to 2011, when their numbers dropped so low that the entire salmon fishery from the northern Oregon border to Mexico was closed in 2008 and 2009 and was severely restricted in 2010.

The coping strategies developed by salmon, dubbed the "portfolio effect," have allowed salmon from California to Alaska to survive for millions of years. Salmon hatcheries may produce more juveniles, but they have been unable to overcome the obstacles and habitat destruction that has accompanied water development in the West.

California's Central Valley: Disappearing Wetlands and Lakes

Another poignant example of the environmental price of water development is the disappearance of enormous expanses of wetlands in California's Central Valley, including three enormous lakes that were located in the southern end of central California. A century ago, California's vast Central Valley would change seasonally as precipitation and river flows changed. This expanse of grassland dotted with oak trees in the dry summer season would be transformed into lush wetlands and marshes during the wet winter season as the rivers, including the Sacramento and San Joaquin, drained the Sierra Nevada and overtopped their natural levees. Large trees—sycamores, willows, cottonwoods, and oaks—once grew along these rivers in abundance. The Central Valley was also home to prolific wildlife: grizzly bears, antelope, Tule elk, salmon, and millions of migratory birds along the Pacific Flyway during the winter, including pelicans, geese, ducks, and cranes.

Although hunting, fishing, and other activities took a heavy toll on the wildlife there throughout the late nineteenth and early twentieth centuries, heavier impacts to wetlands began during the mid- to late nineteenth century as settlers were encouraged to drain and reclaim these "unproductive" wetlands. The catastrophic 1861–62 floods, and large floods that followed, prompted landowners in the Central Valley to begin small-scale land reclamation and flood control measures. But the greatest blow to wetlands of the Central Valley resulted from the era of dam building in the twentieth century, particularly the Central Valley Project in the late 1930s. These water projects, including the building of levees along the Sacramento and San Joaquin rivers and their tributaries, prevented winter and spring floodwaters from spreading out into their natural floodplains, destroying wetlands, shallow lakes, and habitats for fish and wildlife.

The Central Valley once contained four to five million acres of wetlands; by the late twentieth century, only 400,000 acres remained. As the wetlands disappeared, so did the wildlife that depended on them, including the wintering waterfowl—from tens of millions in the 1940s to only about three-and-a-half million by the end of the twentieth century.

Particularly heartbreaking was the fate of the three ancient lakes—Buena Vista, Tulare, and Kern—in the southern end of the Central Valley. These lakes, and the wetlands between them, formed the most extensive contiguous region of wetland in California, covering an estimated 500,000 acres. Tulare was at one time the largest freshwater lake west of the Great Lakes, with four times the surface area of Lake Tahoe. Dating of sediments cored from the Tulare Basin shows that the lake formed over a million years ago, making it one of the oldest lakes in California. These three lakes and their wetlands hosted an incredible diversity of wildlife, with thousands of wild horse, elk, antelope, deer, grizzly and brown bear, wolf, coyote, ocelot, lion, wildcat, beaver, and fox. Birds numbered in the millions, including ducks, swans, marsh wrens, rails, and geese, using the lakes as a stopover point along the Pacific Flyway. Tulare Lake also hosted the southernmost run of chinook salmon in the United States.

The wetlands between these lakes were initially filled in ("reclaimed") following the winter of 1861–62, after the series of severe atmospheric-river storms that generated floods submerging the entire Central Valley (see chapter 2). More serious impacts on the lakes began in the early twentieth century, when the rivers that fed them (the Kern, Tule, Kings, and Kaweah) were diverted to irrigate crops in the Central Valley. What had been the source of water for these great lakes for over a million years was cut off, and the lakes dried up completely. The wildlife that made this region home was lost along with the water. The remaining dry lakebeds were then ploughed and planted with crops, primarily cotton. As a final irony, these crops had to be irrigated with the very river waters that had been diverted from the lakes in the first place. Today, it is difficult to find even the remnants of these ancient lakes within the expanse of crops in the southern San Joaquin Valley.

The Colorado River and Delta

Elsewhere in the West, water development similarly took place without any consideration for the ecological consequences for rivers and their watersheds, floodplains, or deltas. On the Colorado River and its major tributaries, for instance, aquatic ecosystems began declining in the decades after Hoover Dam and fifty-seven other major dams were built on the river and its tributaries starting in the 1930s. Important features of the Colorado River were altered after the construction of these dams, which hold back warm, sediment-laden water, and these sediments then settle out of the still water behind them. At the same time, the deeper parts of the reservoirs become cooler without direct heat from the sun. Thus, the water that is released from the dams to the river downstream is clearer and much cooler than the natural Colorado River water (warm and muddy) to which the native fish species had been accustomed for millennia. All of the 35 species of fish that evolved in the Colorado River—including minnows, chubs, and suckers—suffered from the changing environmental conditions after dams were built.

The Colorado River delta has also been decimated from dam construction and water diversion. The Colorado reaches its delta just south of the Mexican border, where it then empties into the Gulf of California. There, the impacts of damming and water diversions are perhaps the most severe—but least apparent. Before entering the gulf today, only a trickle of the Colorado River remains, and, at times, especially during the dry years, the river no longer reaches its natural delta. Historically, this extensive delta covered about two million acres of woodlands, wetlands, and deserts. In the 1920s, naturalist Aldo Leopold paddled through the "endless green lagoons" (p. 150) of the Colorado River delta to the Gulf of California. Yet today, 93 percent of the delta's wetlands and woodlands have been reduced to a barren, unproductive mudflat.

Major impacts are felt even beyond the delta, in the Gulf of California itself, where the Colorado River once emptied, bringing nutrients, freshwater, and sediments and supporting a prolific marine ecosystem. Freshwater and saltwater mixed in this region for thousands of years, forming rich nursery grounds for fish, shrimp, and waterfowl. But, along with the Colorado River, marine life in the northern Gulf of California has suffered a major decline. Researchers at the University of Arizona have studied changes in the abundance of the Colorado delta clam (*Mulinia coloradoensis*), a species that once flourished at the outlet of the Colorado River in the gulf. These clams provide a measure of ecosystem health, and research shows that, a thousand years ago, these clams were at least twenty times more abundant in the Gulf of California than they are today.

The Sacramento–San Joaquin Delta

The Sacramento–San Joaquin Delta is the hub of California's water distribution system and has been at the center of controversy among environmentalists, farms, fishermen, and cities in the state for four decades. Once a 700,000-acre tidal freshwater marsh, the delta has been gradually converted into a network of channels and islands that limit fish habitat. Moreover, rivers upstream have been impounded behind reservoirs and diverted for agriculture and cities, causing further declines in the delta's habitats. Pumping of water in the southern delta to the San Joaquin Valley farms and Southern California via the California Aqueduct has caused currents in the delta to reverse toward the south, rather than flowing west into the San Francisco Bay and out the Golden Gate. These currents pull small fish toward the southern delta, entraining and killing thousands of endangered delta smelt and juvenile chinook salmon. The delta smelt is endemic to the delta and has been declining since 1990. After 2000, the smelt declined even further as pumping through the delta increased.

In the early 1980s, one proposed solution to these problems was the Peripheral Canal, a forty-three-mile, concrete-lined channel that would have linked the Sacramento River directly to the California Aqueduct. In 1982, however, voters—divided along geographic lines with those in the south voting in favor and those in the north voting against—rejected the construction of this canal. Yet the idea of a canal or tunnel is being reconsidered today, and its proponents argue that it would allow for water exports to the south without the harmful pumping of water in the southern delta. A tunnel, they say, would also protect the delta from seawater intrusion should an earthquake breach the levees protecting this water source. Those opposed to the tunnel are concerned that it would allow for the diversion of too much water away from the delta and the San Francisco Bay estuary, causing increased salinity and further ecosystem declines.

We now understand the impacts that water development has had, and continues to have, on the natural environment and wildlife in the western United States. Policymakers have begun to restore water resources and aquatic habitats, which means that all of us must make water conservation an even higher priority, realizing that we must share this life-giving resource with the natural world. This will be even more imperative in the future, since our water resources may be further taxed as global warming brings changes in climate, sea level, and the hydrologic cycle. This topic is the focus of the next chapter.

THIRTEEN

Future Climate Change and the American West

Many scales of climate change are in fact natural, from the slow tectonic scale, to the fast changes embedded within glacial and interglacial times, to the even more dramatic changes that characterize a switch from glacial to interglacial. So why worry about global warming, which is just one more scale of climate change? The problem is that global warming is essentially off the scale of normal in two ways: the rate at which this climate change is taking place, and how different the "new" climate is compared to what came before.

ANTHONY BARNOSKY, *Heatstroke*

LILACS AND SNOWMELT: WARNING SIGNS

A SURE SIGN OF THE beginning of spring in the American West is the annual bloom of wildflowers: they blanket hillsides and fields, setting them ablaze with color. During the 1990s, Dan Cayan, Susan Kammerdiener, Mike Dettinger, and Dave Peterson, all climate researchers at the U.S. Geological Survey, took an interest in the fragrant lilac. They were looking for plant data that would provide a key to interpreting satellite imagery of the mountain snowpack, a critically important feature of the water supply in the West. They did not start out wanting to know more about flowers; rather, they wanted to know more about snow cover. As it happens, the two are intimately linked.

The idea began in 1957, when Montana's state climatologist Joe Caprio distributed lilac plants (*Syringa vulgaris*) to people across the West, requesting that they send him annual postcard reports with the dates of when the flowers sprouted and bloomed. Lilacs were chosen because they are found throughout the United States, they thrive in all types of soil and climate conditions, and they bloom in response to the increase in spring temperatures.

Four decades later, when Cayan and his fellow climatologists inherited these accumulated data on western spring wildflowers, they found surprising evidence of subtle changes occurring in the West. The researchers pored over the records, teasing out patterns that the casual observer might have easily overlooked. They found that the lilacs had begun blooming throughout the West earlier and earlier over time. By the mid-1990s, the flowers were blooming two weeks earlier than when record-keeping began, signaling that the arrival of spring was coming earlier. The climate researchers compared these data with their stream-flow data across the western United States—from New Mexico to Alaska—and found that mountain streams fed by snowmelt were flowing one to two weeks earlier as well. The researchers set out to investigate why these changes were occurring and what they would mean for water resources in the West.

Why Does Spring Come Earlier?

Initially, Cayan and his colleagues thought that the earlier onset of spring might be associated with a natural climate oscillation, perhaps related to changes in ocean temperature in the eastern Pacific. This was something that the climatologists had seen before in their climate records. But, as the evidence mounted, they realized that this was not just another climate cycle—it was a trend that had started decades before. Its beginnings were tied to the small but significant rise in global temperatures that began in the mid-twentieth century and to the steady rise in heat-trapping greenhouse gases, like carbon dioxide, in the atmosphere. This global temperature increase is reflected in more local records from the American West (see figure 32).

After comparing global temperature data from the mid- to late twentieth century with climate records spanning the past thousand years, most of the world's climate scientists now agree that the period between 1960 and 2006 was substantially warmer than during the previous millennium. Nine of the ten warmest years in the West have occurred since the year 2000, with new records being set every year. This warming is the result of changes in the earth's atmosphere: carbon dioxide levels have risen to 390 parts per million—an increase of approximately 75 parts per million since 1960.

A Changing Storm Track

Satellite observations reveal that, between 1987 and 2006, global precipitation has also increased. Warmer temperatures increase the rate of water

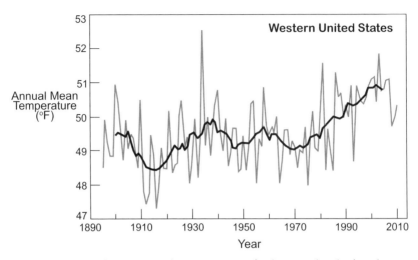

FIGURE 32. Annual mean atmospheric temperature (in degrees Fahrenheit) in the western United States from 1895 to 2010. (Figure courtesy of Kelly Redmond, Desert Research Institute, Nevada.)

evaporation, placing more water as vapor in the atmosphere. Eventually, this additional vapor condenses, resulting in increased precipitation. However, the relationship between warmer temperatures and precipitation is not straightforward. Climate models predict that some areas will experience net losses in moisture because of increased evaporation whereas other areas may experience net gains. Scientists are predicting that the already dry regions, like the American West, will become even drier and that wet regions farther north will become even wetter.

These predictions are increasingly supported by observation. For example, climate models indicate that warming will cause a northward shift of the Aleutian Low and the North Pacific Storm Track, which will intensify precipitation over the Pacific Northwest, leaving the southerly regions, including much of the American West, drier. Researchers at the University of Arizona have observed that, since 1978, the storm track during wintertime (February to April) has been shifting northward, meaning that fewer rain-bearing storms are reaching the southwestern states—that is, Southern California, Arizona, Nevada, Utah, western Colorado, and western New Mexico. These researchers believe the trend will continue into the future if warming continues. Such shifts in the regional water regime may seem subtle in and of themselves, but the impacts are anything but subtle.

The Disappearing Snowpack

The American West depends on snow-bearing winter storms for its water supply. In fact, the region is unique in the contiguous United States in its dependence on the winter snowpack in the high mountains for a natural water reservoir. The snow begins melting in the late spring and continues into the summer, filling streams, lakes, and reservoirs that sustain ecosystems and humans throughout the dry summer months. The snowpack has supported growth in cities and irrigated agriculture. In California, agriculture is an industry that produces 55 percent of the nation's produce, making this semi-arid region an important source of food for the nation. Snowpack provides up to 80 percent of the annual water supply in this region.

A team of climatologists at the Scripps Institution of Oceanography, located near San Diego, predicts that global warming will affect this critical snowpack in several ways. For example, less precipitation will fall as snow as the region warms. For each degree of warming, the snowline moves up the mountainsides by 500 feet. Considering the climate-warming projection for the year 2100 of 5.4°F, the change means snow would fall 1,500 feet higher on the mountainsides; the lower elevations would receive rain. Regions like Northern California that have relatively low mountain elevations in the Sierra Nevada (averaging 7,000 to 8,000 feet) would lose half of their area of snowpack.

Studies have shown that the amount of the snowpack in California has declined by 10 percent during the twentieth century. Climate-modeling predictions suggest that, by the end of the twenty-first century, the snowpack will be reduced by at least 40 percent and perhaps as much as 80 percent. (The amount ultimately depends on just how much warming the region experiences in the future.)

In addition to changes in the volume and location of snow, there will be a change in the timing of snowmelt. The snow that does fall will melt earlier in the spring, rather than during the late spring and early summer when it is so critically needed. The decrease in summer snowmelt will only exacerbate the future drying that will come with warmer conditions during the summer months.

In the 1990s, the U.S. Geological Survey, Department of Water Resources, and other California agencies began continuous monitoring of streamflows, temperature, and chemistry of the Tuolumne and Merced rivers in Yosemite National Park. Changes in stream chemistry provide information about the links between the larger-scale atmospheric circulation over the Sierra Nevada and the hydrology in those mountains. For instance, the amount of silica dissolved in the water provides information about chemical weathering (dissolution) of rocks and minerals in the watershed, and nitrate and nitrite concentrations reflect pollution levels in the atmosphere. Among other results, these studies reveal that the spring snowmelt has indeed been starting earlier.

Earlier snowmelt has propagated impacts downstream as well, such as on water quality and ecosystems in the San Francisco Bay. Monthly monitoring efforts in the bay by the U.S. Geological Survey over decades have shown that earlier snowmelt is causing the salinity of bay waters to rise during the late spring and summer.

Wildfires on the Rise

As the mountain slopes across the great mountain ranges of the West dry, their forests—the pine, Sequoia, and juniper—will be increasingly susceptible to insect infestations, disease, and, inevitably, fire. Diminished winter snowpack, earlier spring snowmelt, and longer, hotter summers will weaken the entire forest ecosystem, leaving the trees fragile and flammable, increasing the number of forest fires across the West.

Residents of the West are all too familiar with this lethal effect of warming and drying. For instance, Southern California residents experienced a catastrophic fire season in 2007 in which close to a million people were evacuated and two thousand homes were destroyed, resulting in $1 billion in damages. For the climatologists at Scripps, this fire season was a painful reminder of the precarious conditions of life in the West. The region had suffered from a lack of adequate precipitation for seven years, including a deep drought in early 2007. Weakened by the dry conditions, millions of acres of spruce, piñon, and ponderosa pine succumbed to bark beetle disease, leaving vast stands of dead trees across the region that acted as kindling during the summer heat waves (see figure 33).

The following year, over two thousand wildfires raged throughout the state of California, scorching 900,000 acres. Subsequently, during the fall of 2008, communities in Los Angeles and Santa Barbara were scorched as hurricane-strength winds roared through the region.

The 2007 and 2008 fires were devastating, but were they a sign of a system out of balance? After all, fire has been a natural phenomenon in California

FIGURE 33. Forest fire in Sedona, northern Arizona, in June 2008. (Photo by B. Lynn Ingram.)

and the West for thousands of years. Climate researchers, including Dan Cayan, wanted to find out whether Southern California was entering into a new fire regime. In a study of twentieth-century wildfires, they found that wildfire frequency did indeed increase over the century, with four times as many fires after the mid-1980s as before.

California is not alone in its plight: fires have been on the rise throughout the American West. Thomas Swetnam, co-director of the Laboratory of Tree-Ring Research at the University of Arizona, sees increased wildfires as one of the first big indicators of climate change in the United States. Warming, reduced precipitation, and earlier spring onset are all linked with larger forest fires. Swetnam notes that these impacts are not probable events in the future; rather, they are happening today.

The Drying Colorado River Basin

In the Southwest, a vast desert region that relies heavily on the Colorado River, similar effects concerning snowpack are predicted. Climate modeling of the Colorado River basin predicts that the snowpack there could decrease by 30 percent as early as 2050. The headwaters of the river lie in the high elevations of Colorado, Wyoming, and New Mexico. The river then flows through the high Colorado Plateau, ultimately bringing precious water to the lower desert states of Nevada, Arizona, and California. Under natural conditions, the river would then drain through its delta into the Gulf of California. Today, all of that water is spoken for and used before it ever reaches the once-thriving delta, as discussed in chapter 12. For decades, every drop has gone to sustaining a growing population and expanding agriculture throughout this arid region.

The stretch of the Colorado River in northern Arizona, between the two largest reservoirs on the river (Lake Powell and Lake Mead), supplies water to over 30 million people in major Southwest cities, including Los Angeles, Phoenix, Las Vegas, and San Diego. Climate models of the region predict that, over the next thirty to fifty years, the Colorado River system will experience 10–30 percent reductions in runoff as the result of climate change. Marine physicist Tim Barnett and climate scientist David Pierce at Scripps have incorporated these predictions into a water budget analysis assessing the impacts of natural climate variability, future climate warming, and water usage on the Colorado River basin. Their results show a net deficit of almost one million acre-feet of water per year in that basin. At this rate, they conclude, there is a 50 percent chance that both Lake Mead and Lake Powell could reach "dead pool," rendering them useless for hydroelectric power generation or useful water storage, by as early as 2021. In the decade between 1999 and 2009, the level of Lake Powell dropped by 60 percent of its "full pool" capacity. The falling reservoir level has exposed enormous white calcium carbonate coatings, rising as high as ten-story skyscrapers on the canyon walls, and these are likely to grow higher in the future (see figure 34).

ECOSYSTEM IMPACTS

As difficult as these changes to our water supply will be on cities and agriculture, we must also consider the profound impacts of altered hydrology on the West's ecosystems. At the University of California, Berkeley, ecologist John Harte has been studying the consequences of global warming in the American West. He has warned that, if all of the snow in mountains of the

FIGURE 34. Lake Powell, in 2009, showing a white calcium carbonate "bathtub ring" exposed after a decade of drought lowered the level of the reservoir to 60 percent of its capacity. (Photo courtesy of U.S. Bureau of Reclamation.)

West were to melt by May, there would not be any water supply during the mid- to late summer, severely affecting the animals and plants that depend on that water.

In an effort to predict the possible ecological consequences of these changes, Harte's research team has been conducting empirical experiments artificially heating wildflowers in a Colorado meadow near Crested Butte since the early 1990s. This research documents the fact that wildflowers bloom earlier when temperatures rise, and it indicates the consequences of that trend. In many cases, wildflowers depend on, and sustain, migrating butterflies that pollinate them. But the pollinators are not arriving earlier: their migration is triggered by day length, not temperature. This has dire consequences for the wildflowers (not to mention for the butterflies as well), threatening an enormous decline in their abundance.

Dozens of plant and animal species are now threatened by climate-induced changes in their food and water sources, habitats, and natural temperature ranges. In prehistoric periods of climatic warming, plants and animals had the time and space to gradually shift their habitats to higher elevations or latitudes. Today, many of these habitats have been reduced to mere remnants—islands in the midst of human development—leaving threatened populations with nowhere to go. Many are already on the verge of extinction from other pressures, such as habitat loss, over-exploitation, and environmental degradation.

A RENEWED DUST BOWL

As the climate warms and dries, the Southwest also faces an increase in atmospheric dust levels from both natural and human causes. Under natural conditions, the desert soils of the Southwest are stabilized by "desert crust"—a mixture of soil, cyanobacteria, and lichens. The cyanobacteria secrete a sticky material that holds soil particles together and retains water and nutrients important for the sparse desert vegetation.

As in the Great Plains of the 1920s, when farmers ploughed the soil too deeply, leaving it vulnerable to wind erosion and leading to the Dust Bowl drought of the 1930s, humans today are disrupting the natural protective surface of desert crust with vehicles, livestock, housing developments, and recreational activities. These effects are exacerbated during droughts (like the decade-long drought between 1999 and 2009) because the bacterial filaments between the soil particles become dry and more fragile, making them more easily crushed. As microbes and lichen decline and the desert crust erodes, the unstable soils are more susceptible to wind erosion, leading to an increase in atmospheric dust.

In lakes near the desert Southwest, researchers have collected cores from lakebeds and analyzed the sediments to calculate the levels of atmospheric dust over the past several centuries. The cores show a six-fold increase in dust after 1850, when livestock were first introduced to the West. The dust levels then declined somewhat after the 1930s, when the government passed the Taylor Grazing Act to reduce overgrazing in the West. But dust levels have been on the rise in recent decades as the human population has grown.

The increased dust levels also have other unforeseen negative impacts on hydrology in the West. For example, as dust settles onto snow in adjacent mountains, it darkens the surface, lowering its reflective properties and causing it to absorb more sunlight, which causes the snow to melt faster, leaving less snow for the dry summer months.

Ironically, even as drought has become the most pressing concern in the American West, climate scientists and hydrologists are warning about increased flood risks throughout the region. Climate researchers Tapash Das, Mike Dettinger, Daniel Cayan, and Hugo Hidalgo have described how, even in a West with reduced precipitation, warming conditions can lead to more frequent flooding in mountain watersheds. Only a few degrees in temperature can determine whether precipitation falls as rain or snow over a watershed. In the mountains, as we have seen, these few degrees translate to a shift of several hundred feet in snow-line elevation. As temperature across the region rises, the total area above the snow line is reduced, with a greater area receiving precipitation as rain. The increase in rain during the short winter months generates runoff that will overwhelm flood control systems and overtop the banks of river channels, leading to floods. A particularly hazardous combination includes early winter storms that drop snow in the mountains followed by early spring storms that bring warm rains in these higher elevations, resulting in rapid and catastrophic snowmelt.

Das's research team has simulated floods on the western slope of the Sierra Nevada, using projected temperature and precipitation patterns and assuming that greenhouse gas emissions will increase throughout the twenty-first century—the so-called "A2" scenario. Their results indicate that, between 2050 and 2100, both the northern and southern Sierra Nevada will experience larger and more frequent floods. The causes include increased frequencies of storms, greater amounts of precipitation, and a greater proportion of precipitation falling as rain rather than as snow.

Mike Dettinger has also independently explored the possible effects that warming in the twenty-first century could have on atmospheric river storms. His analysis shows that, under the A2 greenhouse gas emissions scenario, the nature of atmospheric river storms may change, with increasing frequency of such storms and increases in water vapor transport rates as compared to the historical record. Predicted storm temperatures will be higher, meaning more rain will fall during these events. Dettinger is concerned that these changes will lead to more frequent and larger floods in the future. He predicts a lengthening of the atmospheric river storm season, meaning floods could potentially occur over a longer portion of the year. This will make it difficult for water managers to operate reservoirs, where they are balancing flood control during the wet season with water storage for the dry summer and fall months.

COASTAL IMPACTS

By the year 2050, when the average global temperature is predicted to be about 2°F higher than it was at the turn of the twenty-first century, scientists are forecasting a number of other changes in the West that will directly or indirectly affect the water supply. For instance, current predictions indicate that sea levels along the Pacific coast will rise about thirteen inches or more due to melting ice and thermal expansion resulting from the increased heat absorption by the ocean surface. Rising sea levels will inundate coastal wetlands. The more protected parts of the coastline—bays and estuaries like San Francisco Bay and Puget Sound—will grow larger and saltier with the rising sea, encroaching on the low-lying bordering cities. Around these estuaries, the tidal marshes that had so long been part of the natural history will be squeezed out.

More insidious, though, will be the intrusion of saline water into underground freshwater aquifers that now provide much of the region's drinking and irrigation water. Coastal ground waters are also at risk of becoming saltier as sea levels rise.

In California's Sacramento–San Joaquin Delta, higher sea levels may cause levees to fail, drowning the surrounding low-lying valleys. By 2050, two-thirds of the state's projected 50 million residents may live in the southern Central Valley and Southern California, areas that depend on the freshwater that passes through the delta and is pumped southward through the California Aqueduct. Should the levees fail, two-thirds of the state will be left without a significant amount of their potable water.

INCREASED RISK OF DAM FAILURES

Those who live on floodplains are no longer the only ones at risk of catastrophic floods. The twentieth-century dams hold back vast reservoirs of water, and they have the potential to fail, with catastrophic results. Southern California already experienced a taste of this in 1928, when the St. Francis Dam collapsed, resulting in hundreds of deaths. It was the worst civil engineering disaster of the century in the United States.

Although the twentieth-century era of extensive dam construction (the "hydraulic era," as discussed in chapter 12) essentially came to an end in the 1970s with the rise of the environmental movement, residents of the West will have to contend with the risk of dam failures in the future because aging dams are increasingly stressed by the changing hydrology of the West. As the climate warms, more precipitation will fall as rain and less as snow, and this means more runoff during the winter and more pressure on the reservoirs.

Most of the region's dams were built more than half a century ago—and some are considered disasters waiting to happen. For instance, Glen Canyon Dam on the Colorado River, located on the Arizona-Utah border, began to show signs that its foundation was deteriorating in June 1983. This project had been controversial when it was built in 1956. The resulting reservoir submerged 186 miles of magnificent canyon lands in northern Arizona and southern Utah, including over a hundred side canyons and over two thousand Ancestral Pueblo archaeological sites. After the wet El Niño winter of 1983, unusually large amounts of meltwater entered the reservoir from the watershed. The excess water that poured through the overflow spillways began to turn red, indicating that the spillways were beginning to erode the red sandstone bedrock that supported the dam. James Powell vividly described the unfolding disaster in his book *Dead Pool: Lake Powell, Global Warming, and the Future of Water in the West*:

> Leaks sprang from joints in the outlet works. High pressure popped up manhole covers all over the dam, as if a master magician had levitated them. Everything leaked that could. The water rumbling through the spillways and the vibrating dam produced a cacophony of sound. Standing in the access tunnels in the dam abutments was like being inside a factory in a rainstorm, as the enormous pressure forced water through the porous sandstone. Those approaching the dam from downstream could hear the noise from four miles away. At two miles, large waves stirred the surface of the river and a violent rainstorm fell from the mist emitted from the spillways. Springs spurted from the sandstone walls of Glen Canyon. Closer to the dam the jets from the spillways began to eat into the protective apron that led back to the base of the dam where the generator releases emerge. To one making the trip upstream, that the largest dam disaster in human history might be under way did not seem far-fetched. (p. 14)

Fortunately, the dam remained standing. According to Powell, its collapse would have resulted in a 580-foot wall of water travelling 300 miles downstream, destroying Hoover Dam and Lake Mead, primary sources of water and power for 30 million people in Los Angeles, Phoenix, Tucson, and Las Vegas.

HUMAN-CAUSED OR NATURAL CLIMATE CHANGE?

Climate researcher Tim Barnett and his colleagues have analyzed the climate and hydrology of the American West over the past sixty years, particularly river flow data from the Columbia River, Colorado River, Sacramento River, and San Joaquin River, draining watersheds in the Cascades, northern Rockies, and Sierra Nevada across nine western states. After the research team demonstrated that their global climate model could realistically portray the natural climate in the West, they then ran climate simulations of the West to predict changes in snowpack and river flows.

Using these data and some very long (greater than 800-year) simulations of natural variations (without the human-caused greenhouse gas additions) of climate and river flows from the models, Barnett and his team showed that changes in temperature and snowpack over the last half of the twentieth century were not simply the result of the natural, long-term climatic shifts of the Pacific Decadal Oscillation (PDO). Although the PDO moved to a warmer phase in 1976–77, which was consistent with warmer temperatures and earlier spring snowmelt in the West, it then swung back to a cool phase starting in 1999. Yet this phase shift did not reverse the trends of increasing temperatures and earlier snowmelt. The researchers also showed that volcanic eruptions and solar activity occurring over the past several decades could not have been responsible for changes in snowpack and temperature—in fact, such events would have caused an opposite response: a decrease in temperature and an increase in snowpack. Their conclusion was that only anthropogenic increases in greenhouse gases—that is, human causes—could account for the rising temperature and diminishing snowpack trends they were seeing.

The days of water "on demand" will surely end in the West, replaced by desperate efforts to capture and control what water is left. The Southwest will become a region of extremes as the early spring brings massive flooding and the relentless summer bakes the region dry. While climate scientists have presented the potential risks facing our society, environmental scientists and innovators have explored the various steps we can take to better prepare for a hotter, drier (and sporadically much wetter) future in the West. For example, implementing more effective policies of water conservation and limiting new growth and development in threatened regions will greatly improve the chances of surviving and even flourishing in the future. The next chapter will discuss some of these alternatives as well as the lessons we have learned from the past.

FOURTEEN

What the Past Tells Us about Tomorrow

> If future generations are to remember us with gratitude rather
> than contempt, we must leave them more than the miracles of
> technology. We must leave them a glimpse of the world as it was
> in the beginning, not just after we got through with it.
>
> PRESIDENT LYNDON B. JOHNSON
> *(said while signing the Wilderness Act, 1964)*

AS SHOWN THROUGHOUT THIS BOOK, the American West faces a climatic future that is predicted to become generally warmer and drier, with deeper droughts interspersed with larger and more frequent floods. Scientists believe these shifts may have already begun, since the region is experiencing more extreme weather. Only time will tell, but greenhouse gas–induced warming may be at the root.

Reducing the uncertainty of future climate predictions requires an in-depth understanding of the natural patterns and range of climate, including the fluctuations of conditions experienced over the millennia prior to human-caused greenhouse warming. As discussed in the preceding chapters, a substantial body of research shows that the so-called "normal" range of past climate is enormously broad. Records of long-term climate underscore that, over the past 150 years, nature has dealt the western United States a relatively benign hand. If the decade- to century-long climate extremes written about in this book should revisit the region, modern society will be forced to make dramatic changes in the way we live.

Over the past century, society in the West has followed an unsustainable pattern of water use, leaving us more vulnerable each year as we head into a climatic future we cannot control. The exploding population in the region has been made possible through extensive water development, with dams and aqueducts controlling the distribution of water and buffering the dry years. Nevertheless, we cannot stop droughts from happening. The West is drying, even though water continues to be delivered to our faucets. Key questions probably keep regional water managers awake at night, among them: How bad can it get? How prolonged and severe the droughts? How massive the floods?

The truism that history repeats itself applies not just to the affairs of humankind but to climate as well. The climate history of the past 20,000 years reveals that the American West has been plagued by deep and prolonged droughts, as well as by enormous floods, on a fairly regular basis. Paleoclimate records from across the region, extracted from many sources discussed in this book—tree rings, sediment cores, lake levels, and sediment geochemistry, among others—have been compiled over the years to produce this history. Although some questions remain unanswered, the records have informed scientists of the range of climate conditions the region has experienced, including the frequency and magnitude of extreme climate events. These extreme events far exceed those experienced over the past century and a half. How likely are such events to recur, and are we prepared for these events today?

Droughts

During past periods of climatic warming, such as occurred during the mid-Holocene and again in the late Holocene, conditions throughout the West tended toward increasing dryness, with droughts that could stretch out for centuries. Research from California and the Great Basin, for example, shows that the mid-Holocene was dry for nearly 1,400 years. Humans occupied the American West during this time, and the archaeological record shows that the mid-Holocene warming coincided with mass migrations of native populations from the desert interiors of the West to the coastal regions. This long episode of warmer and drier climate was followed by the Neoglacial—a period of cooler and wetter climate.

In the late Holocene, beginning approximately 1,800 years ago, drying conditions returned, peaking during a particularly dry period from AD 900 to 1400 (the so-called Medieval Climate Anomaly). Prolonged droughts were a fixture during this drier millennium, as were intermittent floods. Combined, these extremes brought down flourishing civilizations that, the archaeological records show, all but disappeared as their agricultural bases withered. The scattered native peoples that were able to persist during these prolonged dry spells in the American West were also hard-hit by extreme floods, so they had to develop coping strategies. One key to their success was an intimate knowledge of their environment and its resources. Another was the size of their population: they survived by living in small, flexible groups, able to mobilize rapidly and relocate as conditions dictated.

The hardships suffered by these early human inhabitants of the West provide important lessons. For instance, during extended periods of abundant moisture, societies that grew rapidly were left vulnerable to collapse when the climate inevitably turned dry again. Modern society in the West has followed a similar path: after a century of fairly abundant moisture, the population in this region has exploded. Modern engineering has allowed the exploitation of all available water sources for human use, and water policy has favored water development for power, cities, and farms over sustainability of the environment and ecosystems. These policies have allowed populations to grow right up to, and perhaps beyond, the limits that this region can support, leaving it vulnerable should drier conditions return. In this respect, today's society in the West resembles the Ancestral Pueblo of the twelfth century—both living beyond their means.

Paleoclimate records also reveal that the lack of prolonged droughts during the past 150 years is an aberration. In the past two millennia, such droughts, some of which lasted several decades, occurred at intervals of every fifty to ninety years. The longest droughts experienced by the West over the past century—lasting about six years—are meager by comparison. Certainly, however, as discussed in chapters 1 and 3, these relatively short droughts brought extreme hardship to the region.

Climate scientists have discovered a new complication in the West. In addition to that region's variable climate, scientists agree that a warming trend has been superimposed on it—a trend that began a few decades ago and is likely caused by a rapid increase in carbon dioxide (and other greenhouse gases) in the atmosphere. The American West appears to be facing a potential climatic double-whammy: a cyclical return of the drier conditions and a new greenhouse gas–induced warming. Although we do not know exactly what kind of weather this combination will produce, climate models indicate that global warming will compound natural climate cycles and that a warming world will likely make the extreme events, particularly floods and droughts, even larger and more frequent.

Floods

The paleoclimate records suggest that the West has experienced far worse floods, at regular intervals, than the region has seen over the past 150 years. The 1861–62 flood, for instance, was the result of a series of large storms that flowed one after another across the Pacific in what climatologists describe as an atmospheric river. These storms slammed the West Coast—from the Mexican border to Canada—for 43 days, bringing triple the average amount of rainfall to the region and leading to unprecedented flooding. Based on the relatively short historic record, water planners have estimated that a flood of this magnitude may recur perhaps every thousand years. However, paleoclimate records spanning the past two thousand years indicate that floods of this magnitude, or perhaps larger, have recurred every one to two centuries.

The largest of these floods occurred during the Little Ice Age, four hundred years ago. This event predated the history of modern California, yet it is interesting to note that Native American residents in 1861–62 appeared to recognize the early warning signs of these storms and the subsequent floods, and they quickly moved out of the floodplains and up to higher ground. Their long history in the region had given them a deeper understanding that, though the floodplains could provide abundant rich soils and resources, they could only be safely occupied seasonally. The Native Americans knew better than to build permanent settlements in California's Central Valley or delta.

FUTURE FLOOD RISKS

California's Central Valley

Many regions in the western states are vulnerable to flooding today, despite decades of construction of flood control infrastructure. Modern society has promoted growth, with cities sprawling out onto the floodplains, deltas, and other low-lying areas. Many of these regions are protected by aging levees that have been shown to fail under larger flood events.

Other human activities have added to the flood risk. Take, for example, California's Central Valley: viewed from above, it looks like a giant bathtub—up to 400 miles long and 70 miles wide—with the San Francisco Bay estuary and delta as the drain. Most of California's agriculture, and much of its future population growth, will occur there. Early accounts from California history describe winter floods, like the 1861–62 event, that turned the Central Valley into an inland sea. Floods have been a natural feature of the Central Valley for millennia, which is why the soils are so rich and fertile. However, flooding on this scale does not mix well with the interests of a modern agricultural system that depends on regularity and controlled water supplies, or with the inland cities that are now replacing them.

Over the past century or so, Californians have prevented these catastrophic floods through massive hydro-engineering projects. Water managers have essentially taken nearly full control of the timing and amount of water that reaches the fertile lands within the Central Valley. The consequence of this transformation of the natural hydrology, combined with groundwater pumping, soil compaction, and the draining of wetlands, has been that much of this vast region has sunk by several feet and, in some areas, by as much as 28 feet.

Today, this sunken region may be more susceptible to flooding than it was in 1861–62. Moreover, the population of California has grown from 300,000 in 1861 to over 37 million in 2012, a number that is expected to double by the year 2050. Room has to be made for this swelling population in the future, and people have been finding that space in the Central Valley. Increasingly, the region has been providing more affordable housing, with former agricultural lands in the Central Valley floodplains being converted to housing developments. Should floodwaters fill the Central Valley, they will submerge not just small towns, marshes, and farmlands but also major population centers in ten to thirty feet of water. In 2009, over six million people lived in the Central Valley, many in densely populated cities like Modesto, Fresno, Stockton, and Sacramento, all of which face potential catastrophe should storms and floods like the ones in 1861–62 strike again. The recent trend in urban sprawl seems to dance on the edge of disaster: flooded fields are one thing, but flooded homes are quite another.

The long history of megafloods in the Central Valley is virtually unknown to most residents today. Many new cities and housing developments are protected by a relatively primitive network of levees—some 6,000 miles of levees in all—that line the rivers. These levees are fragile and they are the last—and in some cases only—line of defense against the potential floods spilling out from the Sacramento and San Joaquin rivers. The fragility of California's levees has repeatedly been exposed in recent years. During the winter of 1986, floodwaters overwhelmed Folsom Dam on the American River and almost flooded Sacramento; in 1997, levees broke throughout the Central Valley, causing the largest evacuation in California's history; and in 2006, multiple levee breaks and massive flooding in the Central Valley made national news.

The Delta

The Sacramento–San Joaquin Delta is located inland of San Francisco Bay on the western edge of the Central Valley, at the confluence of the Sacramento and San Joaquin rivers. These rivers drain a watershed covering almost half the area of California, including the southern Cascade range to the north, the western flank of the Sierra Nevada, and the eastern flank of the Coast Ranges. Wet winters bring huge volumes of water to the delta—a region now considered a disaster waiting to happen.

The delta was once a 700,000-acre network of wetlands and channels, with seasonally shifting river inflows, and host to millions of migrating birds and fish as well as animals such as elk and grizzlies. During the Gold Rush, however, the delta's natural wetlands were drained and converted to farmland to feed the growing population in San Francisco. Islands in the delta were formed, protected by levees built by farmers, Chinese laborers, and miners, using wheelbarrows, picks, and shovels. These earthen levees confined river flows to the channels, so that the wetlands on the newly isolated islands no longer received sediments or nutrients.

Over time, the sediment debris on the wetlands that had accumulated over centuries dried, oxidized, and decomposed. Winds blew the dried peats away, causing the islands to subside by up to twenty-five feet below sea level. The levees separate the channeled waterways through the delta from these now-sunken "islands." In recent years, over half a million homes have been built on these islands in the delta, yet many of the new occupants are unaware of the substantial flood risks. As the levees continue to age, they are becoming more prone to failure during floods and even earthquakes. In 2004, a 350-foot section of a levee ten miles west of Stockton collapsed, flooding a 12,000-acre island called the Upper and Lower Jones Tracts (see figure 35). As discussed in chapter 12, a breach in a levee in the delta also has the potential of allowing seawater to intrude into the freshwater supply that is pumped to residents in Southern California as well as to farmers in the San Joaquin Valley.

California's capital city of Sacramento, located just north of the delta along the Sacramento River and just downstream of the Folsom Reservoir, is also at the mercy of levees that are over a century old. This city, which was under ten to twenty feet of water in the 1861–62 flood written about in chapter 2, has the lowest level of flood protection for an urban area anywhere in the United States, despite its dramatic flood history. With flood protection well below

FIGURE 35. Sacramento–San Joaquin Delta levee break in June 2004, which flooded a 12,000-acre island in the delta. (Photo courtesy of the California Department of Water Resources.)

that of New Orleans, a large flood in the Sacramento region could result in tens of billions of dollars in damage and put thousands of lives at risk.

AN ATMOSPHERIC RIVER "SUPERSTORM"

Recent research shows that atmospheric river storms have produced the largest historical floods along the Pacific Coast states of California, Oregon, and Washington. As the global climate warms over the next century, larger and warmer atmospheric river storms (and extended storm seasons) are predicted. From a flood-risk perspective, this is bad news.

The U.S. Geological Survey has published a new emergency-preparedness scenario called "ARkStorm" (standing for Atmospheric River 1000 Storm). This scenario assesses the extent of damage that would result if a hypothetical series of atmospheric river storms, analogous to those that struck in 1861–62, slammed the West Coast today. The ARkStorm scenario shows that flood control systems would be overwhelmed, and Sacramento, as well as the expanding cities and populations in California's Central Valley, delta, and other low-lying regions, would be almost entirely underwater in a "Hurricane Katrina–like" disaster. The ARkStorm scenario also indicates that hurricane-force winds would strike the coast, in some places reaching 125 miles per hour. As slopes and hillsides become quickly saturated, the ensuing landslides and mudflows would cause widespread damage to roads and property. This scenario suggests that one-quarter of all homes in California would be destroyed and an estimated 1.5 million residents would be forced to evacuate the Central Valley and the San Francisco Bay delta.

The total cost to California in this scenario, including business interruption and property repair costs, could reach $725 billion, which is three times the estimated cost of a large Southern California earthquake. The U.S. Geological Survey considers this extreme storm scenario to be the next disaster waiting to happen—even more catastrophic than the next big earthquake predicted to strike the region. The ARkStorm team has raised public policy questions, including whether to fund emergency preparedness and atmospheric river–prediction studies, and it recommends that steps to mitigate the impacts of this scenario should begin immediately to prevent catastrophic loss of life and property.

We now have the ability, as a society, to understand climate and forecast weather. We also have deeper insights into the impacts of extreme climate events on our world. It is time for individuals and society to change our behavior in order to mitigate the effects of these events. One important step would be to allow our floodplains, marshes, and deltas to resume their natural functions. Each year, the oceans rise higher, and some scientists suggest that the best protection from the dangerous combination of rising sea levels and extreme storms is what nature has already provided but we have nearly eliminated: tidal wetlands.

In California before the Gold Rush, for example, the San Francisco Bay delta was surrounded by 849 square miles of tidal wetlands. Not only did these wetlands provide unique habitats for millions of migrating and resident birds, mammals, and fish, but the lush vegetation also slowed storm-driven waves, dissipating their energy. Today, only about 48 square miles of the wetlands remain, and these remnants are among the most threatened ecosystems in the state. An urgent question is whether these marshes can keep pace with the rising sea. The sediments delivered to the marshes may prove inadequate to keep the surface above water, and the marshes may no longer have the option to retreat landward, as they are now completely surrounded by urban sprawl.

Most water experts urge policymakers in the West to begin taking action immediately to prepare for both ends of the climate spectrum facing the region: drier conditions and decreased water availability interspersed with longer storm seasons leading to larger and more frequent floods. Although it is not yet clear to what degree the drought conditions in the West over the past decade have been the result of natural or of human causes, climate experts are recommending that society begin to prepare early for the coming changes.

Jonathan Overpeck and Bradley Udall at the University of Arizona advise scientists and policymakers who are currently discussing strategies for adapting to climate change that they need to include the possibility that the West could enter into another megadrought like the ones discussed in earlier chapters of this book. Their "no regrets" approach includes adapting to overall reductions in water and preparing to live in a landscape that may be transformed by the other impacts of climate change, including sea-level rise, greater flooding, more frequent wildfires, and ecosystem shifts.

This "no regrets" approach is wise and conservative. However, it is difficult to convince a society about the need to prepare for risks when the most recent large floods have become a distant memory. Flood memory half-life, suppressing individual and public memories of recent flooding, has been demonstrated by flood insurance coverage behavior. For instance, the rates for flood coverage increased immediately after the 1997 floods but have been declining since then.

Similarly, it can be a tough sell to convince people to adopt a "no regrets" approach to water conservation when water still flows abundantly and cheaply through our taps. Water is uniquely vulnerable to overuse. It falls freely from the sky, giving a false sense of abundance. Water policy in the West has made it a resource that is easy to seize and exploit by those with the power and will to do so. As yet there is little incentive to conserve water for the common good, despite scores of ingenious water conservation proposals. Society in the West lacks any sense of urgency concerning the growing scarcity of this life-sustaining resource. People in this region are oddly estranged from their natural surroundings—living in human-created oases of concrete, manicured lawns, and air-conditioned homes. Few are aware of the origins of the water flowing out of their taps, which is perhaps due, in part, to the complexity of the region's modern, highly engineered hydrology.

Peter Gleick and Matthew Heberger at the Pacific Institute in Oakland, California, view the decade-long drought in Australia that ended in 2010 as a "wake-up call" for the American West. The Australian drought, the worst in that country's recorded history, had exceptionally low river flows, agricultural collapse, catastrophic fires, and blinding dust storms. The drought-prone West has many lessons to learn from Australia, a similarly semiarid to arid region.

Gleick and Heberger point to some of the strategies adopted by Australians during their drought, including steps that individuals can take to conserve, such as converting to water-efficient washing machines, toilets, showerheads, and faucets, all of which can save an enormous amount of water per year. Recycling and reusing treated wastewater (or gray water) has also been successfully adopted by Australians, including treatment of this water to a level safe enough for drinking. These strategies led to a 37 percent decline in water used per person between 2002 and 2008 in Australia.

Other strategies for water conservation include rainwater harvesting systems that allow individual homes or office buildings to capture rain that falls on roofs and transfer the water to above- or below-ground cisterns or storage tanks for use at a later time. The construction of desalination plants may also be a partial solution along the coast. Australia spent $13.2 billion to build desalination plants in its five largest cities in 2011 in preparation for future droughts.

REDUCING OUR WATER FOOTPRINT

Another important step in water conservation is to improve societal awareness of the "embedded," or virtual, water in foods and products. Dubbed our "water footprint," ecologists and economists have recently defined this concept so that people living in the western United States can learn how much water they actually consume in their daily life. The water footprint is analogous to the more familiar carbon footprint in providing an accounting of the total amount of water used to produce a crop or product. Keeping track of our water footprint will ultimately help individuals, companies, regions, and even countries practice more intelligent water usage.

The water footprint of beef is a telling example for the American West. This region celebrates wide-open spaces and the iconic cowboys on horseback, driving cattle. One of its signature products—beef—is the most water-intensive food produced anywhere. The water footprint of beef includes water used to grow feed grain, water used to produce the antibiotics administered to cattle during their three-year average lifespan until slaughter, and water used in processing the meat. The water required to grow the feed grain alone is about 4,042 gallons per animal. In addition, a cow drinks on average 6,340 gallons of water in three years. Added together, the amount of water used to produce a single, six-ounce serving of steak is 2,600 gallons. Dairy products are equally water-intensive for the same reasons. Forty-nine gallons of water are used to produce a single eight-ounce glass of milk.

These numbers can radically alter how we compute actual water usage. On average, the direct water usage per day—water used for drinking, cooking, bathing, washing clothes, sanitation, and landscaping—is currently about 1,430 gallons per person in the American West. If the indirect, or embedded, costs are included, the average water footprint per person per day jumps to 4,110 gallons per day, or 1.5 million gallons per year—enough water to submerge a football field to a depth of four feet.

The water footprint of an average resident of the American West is significantly greater than that of a resident of the eastern United States. Higher overall rainfall in the East, including summer rains, reduces the need for irrigation in landscaping and agriculture. In contrast, virtually all crops grown in the arid West, and most landscaping, require irrigation, which increasingly comes from ancient aquifers located beneath the ground. These concepts were written about some twenty years ago by Marc Reisner in *Cadillac Desert:*

> The West's real crisis is one of inertia, of will, and of myth. As Wallace Stegner wrote, somehow the cow and the cowboy and the irrigated field came to symbolize the region, instead of the bison and the salmon and the antelope that once abounded here. But they might be driving bison, in reasonable numbers, instead of cows, and raising them, for the most part, on unirrigated land—which bison tolerate far better than cows. In a West that once and for all made sense, you might import a lot more meat and dairy products from states where they are raised on rain, rather than dream of importing those states' rain. (pp. 514–15)

The western water footprint expands far beyond its geographic boundaries, since half of the nation's produce is grown in the West. Water policy experts are now asking whether it makes sense for California and the West to be growing water-intensive crops—such as cattle feed, rice, and cotton—for the rest of the country.

Important successes over the past several decades illustrate the difference that ordinary citizens can make in reversing the environmental and ecological damage resulting from modern water development. The restoration of Mono Lake in the 1980s and 1990s after four decades of water diversion by the city of Los Angeles is a powerful example. Between 1941 and 1981, diversions of inflowing streams lowered Mono's lake level by forty-four feet, reducing the lake's volume by half and threatening a fate similar to that of Tulare Lake, another ancient and biologically important lake in the southern San Joaquin Valley, now gone due to diversions (see figure 36). The fascinating tufa towers that make Mono Lake famous were originally formed underwater. As the lake level dropped, these formations were exposed (see figure 37). Mono, a saltwater lake, became twice as salty after the diversion of its inflowing streams reduced the lake's volume by half. These diversions also affected ecosystems and vegetation along Mono's inflowing streams, leading to a steep decline in trout and other aquatic organisms. The exposed lakebed around the edges of the lake dried and was entrained by winds, causing unhealthy levels of dust in Mono Basin—severe enough to violate federal air quality standards.

The Mono Lake Committee was formed in response to this emergency. This organization of 20,000 concerned citizens worked with the National Audubon Society to research the impacts of water diversions on the lake's unique ecosystem, including trillions of brine shrimp and alkali flies, algae, and millions of migratory birds that rely on the lake's resources each winter for feeding.

The coalition finally sued the Los Angeles Department of Water and Power, charging that it was in violation of state fish and game codes. In a historic decision in 1983, the California Superior Court ruled that the public trust doctrine applies to Mono Lake. The essence of this doctrine is that public waterways (including streams, wetlands, tidelands, and lakes) should be protected for all citizens for navigation, commerce, fishing, recreation, and aesthetic values, including wildlife. The court ruled that the state is not bound by past decisions on water usage, particularly those that do not accord with current knowledge or needs to protect public trust uses of Mono Lake.

Because the lake's ecosystem was threatened, Los Angeles was ordered to restore 60,000 acre-feet of water per year—enough to allow the lake to stabilize and begin to regain its former size. Although today the lake is still twenty-five feet lower than pre-diversion levels in 1941, its ecosystem is rebounding

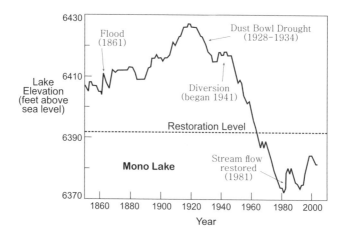

FIGURE 36. Level of Mono Lake (feet above sea level) for the period 1850 to 2011. Diversion of inflowing streams began in 1941. Diversions were reduced in 1981, allowing the lake level to rise and the ecosystem to begin to recover. (Data from Mono Lake Commission website, plotted by B. Lynn Ingram.)

FIGURE 37. Mono Lake, showing calcium carbonate "tufa tower" formations that originally formed beneath the lake but are now exposed after the water level dropped. The eastern flank of the Sierra Nevada range is shown in the background. (Photo by D. J. DePaolo.)

and its salinity levels have gone down. Water has also been returned to the Owens River, allowing it to flow again for the first time in fifty years.

This case profoundly affected California's water policy, making it impossible for water administrators to neglect the needs of ecosystems in water policy decisions. The story of Mono Lake is considered a shining example of what is possible in environmental restoration of aquatic habitats, including rivers, lakes, deltas, estuaries, and marshes, that have suffered under water development.

The success at Mono Lake provides hope that perhaps other regions of the West decimated by water development will be revived in the future. For instance, biologists studying the Colorado River delta have noted that, after unusually wet winters such as the El Niño years of 1983, 1988, 1993, and 1997, winter rains and spring snowmelt filled the reservoirs on the Colorado River to the brim, forcing dam operators to release excess water. This relatively small amount of excess river water reached the delta and was sufficient enough to revive the habitat of a small region, perhaps 150,000 acres. Biologist Ed Glenn and his colleagues at the University of Arizona have speculated that these small releases mimic the spring floods that used to reach the delta before the dams were built. He proposes that 1 percent of Colorado River water be allocated for the delta every year in order to restore and maintain many of its key habitats.

Tulare Lake in California's southern San Joaquin Basin is another candidate for restoration. Its inflowing rivers were diverted and the lakebed was converted to cotton fields, but, during very wet years, a small part of the lake basin floods. The nonprofit Tulare Basin Wildlife Partners has been preserving the one remaining creek that flows to the lake from the Sierra Nevada, and the group has begun small-scale restoration of wetlands around the lake.

DAM REMOVAL

Policymakers are analyzing other ways to reverse the collateral damage to our ecosystems from water development, including dam removal. In the arid West, colossal dams were once thought to be critical to prosperity. Today, there are alternatives that can achieve the functions provided by these dams, without the damage. Wetland restoration and riparian buffers can provide flood control. Limiting development in the floodplains can decrease risks from floods. Dam removal becomes a more reasonable proposal when we consider that dams have finite life spans, determined in part by the rate at which they fill with sediment. This has been known at least since the mid-1900s, when ecologist and environmentalist Aldo Leopold noted: "We build these dams to store water, and mortgage our irrigated valleys and our industries to pay for them, but every year they store a little less water and a little more mud. Reclamation, which should be for all time, thus becomes in part the source of a merely temporary prosperity" (p. 179).

This temporary prosperity comes at an enormous price. When the costs of a dam (environmental, cultural, or safety) outweigh the benefits (flood control, hydropower, water storage, and irrigation), we should consider removing it. Dam removal can be done successfully: in the United States, at least 600 dams have been removed since 1912; the majority of them were removed after 1980.

Along the Pacific Coast, dams are being proposed for removal in California, Oregon, and Washington in order to restore salmon spawning habitat. For instance, four dams on the Klamath River at the California-Oregon border are slated for removal to restore salmon habitat after Northern California Indian tribes filed lawsuits to restore their fishing rights. These tribes had relied on salmon from the Klamath for food as well as cultural and spiritual sustenance for millennia. After dam construction, salmon populations plummeted on a river that once supported the third largest salmon run in the West.

TOWARD A NEW POLICY OF BALANCE AND COOPERATION

The past century has pitted ecosystems in the West against an economic imperative of growth by an ambitious western society. Yet with each drought or flood, the costs mount—both to the ecosystems that are collapsing and to the economy that reels with each disaster. An alternative paradigm for the future would consider the relationship between sustainable water use and the delicate balance of life within an environment that has evolved over many millennia. Human society must find its place as part of the delicate balance, not apart from it. Achieving this balance will be both increasingly important and increasingly difficult in the future should the region's climate, as predicted, become drier and punctuated by more floods. Now is the time for policymakers and residents to move toward a new, overarching policy of collaboration, to act together for the collective good and the survival of the region.

One example of collaboration can be found in Northern California's San Francisco Bay delta, which is currently a focal point of battles between water users and a collapsing ecosystem. The delta, which provides water for 27 million Californians (mostly in the southern part of the state) and millions of acres of farmland in the arid San Joaquin Valley, is the hub of the state's water system. Many of the lessons that are being learned the hard way in the delta should help other similar water conflicts throughout the West.

As discussed in chapter 12, water development over the past century—resulting in changes in total river inflows and timing, salinity, water temperature, suspended sediment loads, contaminants, and nutrient levels—has greatly affected aquatic ecosystems and native fish habitats in the San Francisco Bay and delta. These impacts include population losses resulting in several native fish species (such as chinook salmon and delta smelt) being listed as threatened and endangered. In recent years, efforts to protect endangered fish have restricted the timing and amount of water diverted from the delta for the San Joaquin Valley and Southern California cities, leading California's legislature to enact the Delta Reform Act in 2009. This act mandates that water planning and policy going forward in California now include two "co-equal goals": protecting and rehabilitating the delta ecosystem, and supplying reliable water to the state.

Two major reports on sustainable water and environmental management in the delta were published in 2012 and 2011 by the National Research Council (NRC) and the nonprofit Public Policy Institute of California (PPIC), respectively. According to the former, an important first step toward a more sustainable water future is acknowledging that water scarcity is now a way of life. This recognition would allow water policies addressing this scarcity to be drafted and implemented. In California, as in much of the West, most of the surface waters have been allocated or even over-allocated; there is not enough water to reliably meet all desired uses and needs. Water scarcity has increased over the past century with the growth of population and irrigated agriculture. As discussed in earlier chapters of this book, the natural climate in the West is highly variable, and the past century of relatively wet climate may shift back to a drier phase. In addition, predictions are that the West will be even drier in the future as the climate warms. Coupled with continued population growth, the region will most certainly experience more severe and long-term water scarcity.

The NRC report notes that water scarcity has only been addressed in the past during periods of drought, using water rationing as the main coping strategy. Going forward, it recommends a long-term sustainable policy that incorporates water scarcity into water planning and management, such as pricing water to reflect its long-term scarcity and improving water conservation, efficiency, and productivity. The California legislature set new water conservation standards in 2009, mandating a 20 percent decrease in water usage per person by the end of 2020 in the urban sector. There are numerous ways of achieving this goal, including more water-efficient appliances, drought-tolerant landscaping, water recycling, and rainwater harvesting. However, agricultural water usage accounts for 77 percent of water usage in the state, yet it has not been held to a similar standard of water reduction. Many water-efficiency approaches that are now available to agriculture were discussed in the NRC report, such as using a tiered pricing structure, adjusting the timing and amount of irrigation to maintain specific soil moisture requirements of crops, replacing irrigation systems with closed-conduit systems (particularly drip or trickle systems), and recycling excess irrigation water.

The PPIC has further insights in its report, which notes that water is both a public good, providing critical environmental services, and a commodity, with a market value, and ultimately must be managed in a way that fulfills both of these roles. A balance must be struck between environmental and human needs in the West, and both water scarcity and environmental costs should be reflected in its price. Increasing water rates, for instance, by using tiered rates that increase with greater water usage would lead to greater conservation and increased water efficiency by urban and agricultural consumers. Higher-priced water would also encourage a shift from wasteful, lower-value, water-intensive uses (including the growing of crops such as cotton, rice, and alfalfa) to higher-value agricultural, urban, and environmental uses. Another strategy suggested by the report is the enactment of a "public goods" charge for major water users to generate funding for water management, infrastructure improvement, and environmental protection and restoration.

In addition to conservation of water through pricing, the PPIC report considers water supply sources and proposes using many different sources of water as a key strategy. This "portfolio approach" includes nontraditional sources like recycled wastewater for urban use and recycled irrigation water for agricultural use. Another key strategy discussed in the report is massive water storage in underground aquifers, which could serve as a water bank for drought years, particularly during multiyear droughts when surface reservoirs shrink. Groundwater storage would also buffer future shifts in the water supply when winter floods, caused by the shift from snow to rain, become larger and more frequent: an increase in the amount of water stored underground would free up surface reservoirs for flood control. Underground storage might also be less costly and more environmentally friendly than building more surface reservoirs.

Peter Gleick and his colleagues at the Pacific Institute point out that the twentieth-century approach to water policy was in creating "new" water for human use by building water infrastructure—such as dams and aqueducts—for water storage and distribution. (For further discussion, see chapter 12 in this book.) They argue that it is time to create new water through conservation and improved water-efficiency measures. Moreover, improving the efficiency of water use can have additional benefits, including better water quality, healthier aquatic ecosystems, greater reliability of the water supply during droughts, decreased energy demands, reduced water infrastructure costs, and higher crop productivity. Gleick's team also points out that reducing unproductive uses of water is another important strategy in coping with water scarcity. For instance, improved technology and water management can reduce the amount of water evaporated from impermeable urban surfaces and irrigated fields or blown off large sprinklers used by farms. The team predicts that improvements in water stewardship practices in the agricultural sector, including reducing unproductive water loss, will go a long way toward helping the state cope with extended droughts and long-term water scarcity in the future.

Water policy experts and climatologists have also begun working together to integrate water management, flood protection, and restoration of aquatic ecosystems. For instance, Michael Dettinger and his colleagues at the U.S. Geological Survey have shown that a small number of very large storms provide 30–50 percent of California's annual water supply. (Atmospheric river storms and the devastating floods they produce are described in chapters 2, 4, and 10 above.) They point out that much of the water from these storms is lost to human and ecosystem use during flooding because managers are forced to release water from over-full reservoirs. The team advises that state resource planners coordinate water management with flood control efforts, allowing the restoration of some floodplains to their natural function. This approach would increase the capacity of reservoirs without hampering flood control.

Today, scientists have a much greater understanding of the causes and consequences of the West's climate than at any time in the past. We have also gained a broader perspective on the natural climate fluctuations in the region over the past millennia. The long-term history of natural climate patterns described in the pages of this book reveals that extreme flood and drought events struck the West at regular intervals. In addition to the recurrence of these extreme events, a future warmer climate brought on by human activities will bring drier conditions, interspersed with floods, to the region. The time for the West to prepare for this drier and more extreme future climate is now, rather than waiting until we are in the midst of a crisis and attempting to respond.

Fortunately, many strategies and options are available to the region. A more sustainable water future in the West would include linking urban growth with water supply and availability. Reconciling human and environmental uses of water, integrating management of water supplies, water quality, and flood protection, and managing water more flexibly and responsibly lie at the core of a more sustainable future. Continued research efforts toward forecasting major flood-producing storms as well as wet and dry periods associated with El Niño and La Niña events and other ocean-atmosphere oscillations (such as the PDO) will help with prediction of these extreme events and implementation of warning systems in flood-prone regions such as California's Central Valley and Sacramento delta.

Society as a whole needs to be educated about water's vital importance, its scarcity, and what we can do to help. An important first step would include increasing awareness about both the direct and the indirect water usage in our daily lives—our water footprint. All current residents in the American West would also benefit from reflecting on the native cultures that came before us, that lived more in tune with their environment and understood by necessity the delicate balance between consumption and conservation. Today, our society would do well to learn about this balance if we want to survive and thrive in the future. We should be neither at the mercy of the natural world nor at odds with it. Today, we can choose to reimagine our place in the delicate web of life in the West—by using sophisticated engineering and technology to devise ingenious solutions and by seeing beyond our desire to consume to a future where the needs of a healthy environment are also met. We may find that we have all we need to preserve our quality of life as well as to protect and preserve the natural environment around us for generations to come.

BIBLIOGRAPHY

PREFACE

Byrne, R., B. L. Ingram, S. Starratt, F. Malamud-Roam, J. N. Collins, and M. E. Conrad. 2001. Carbon isotope, diatom, and pollen evidence for late Holocene salinity change in a brackish marsh in the San Francisco estuary. *Quaternary Research* 55: 66–76.

California Department of Water Resources. 1978. *The 1976–1977 California Drought— A Review.* California Resources Agency, Department of Water Resources.

Cayan, D. R., K. T. Redmond, and L. G. Riddle. 1999. ENSO and hydrologic extremes in the western United States. *Journal of Climate* 12: 2881–93.

Gleick, P. H., and L. Nash. 1991. The societal and environmental costs of the continuing California drought. Pacific Institute for Studies in Development, Environment, and Security Research Report.

Ingram, B. L., and D. DePaolo. 1993. A 4300 yr strontium isotope record of estuarine paleosalinity in San Francisco Bay, CA. *Earth and Planetary Science Letters* 119: 103–19.

Ingram, B. L., and D. Sloan. 1992. Strontium isotopic composition of estuarine sediments as paleosalinity-paleoclimate indicator. *Science* 255: 68–72.

Peterson, D., D. Cayan, J. DiLeo, M. Noble, and M. Dettinger. 1995. The role of climate in estuarine variability. *American Scientist* 83: 58–67.

Reisner, M. 1993. *Cadillac Desert: The American West and Its Disappearing Water.* New York: Penguin Press.

Serle, Percival, ed. 1949. Mackellar, Dorothea. In *Dictionary of Australian Biography.* Sydney: Aungus and Robertson.

INTRODUCTION

Carle, David. 2004. *Introduction to Water in California.* Berkeley: University of California Press.

Cayan, D. R., K. T. Redmond, and L. G. Riddle. 1999. ENSO and hydrologic extremes in the western United States. *Journal of Climate* 12: 2881–93.

Frost, R. 1920. "Fire and Ice" in the selection "A Group of Poems" by Robert Frost. First published in *Harper's Magazine*, December issue, p. 67.

Hastings, L. W. (1845) 1994. *The Emigrants' Guide to Oregon and California*. Facsimile edition. Bedford, Mass.: Applewood Books.

Manly, W. L. (1894) 2004. *Death Valley in '49: Important Chapter of California Pioneer History*. San Jose, Calif.: Pacific Tree and Vine Co. Project Gutenberg Ebook, www.gutenberg.org/files/12236.

Powell, J. L. 2008. *Dead Pool: Lake Powell, Global Warming, and the Future of Water in the West*. Berkeley: University of California Press.

Powell, J. W. (1879) 1983. *Report on the Lands of the Arid Region of the United States, with a More Detailed Account of the Lands of Utah*. Facsimile of the 1879 edition. Boston, Mass.: Harvard Common Press.

Reisner, M. 1993. *Cadillac Desert: The American West and Its Disappearing Water*. New York: Penguin Press.

Wilber, C. D. (1881) 2000. *The Great Valleys and Prairies of Nebraska and the Northwest*. Reprinted in *Wilber Republican*, Aug. 2.

PART I. FLOODS AND DROUGHTS IN LIVING MEMORY

Alley, R. 2000. *The Two-Mile Time Machine: Ice Cores, Abrupt Climate Change, and Our Future*. Princeton, N.J.: Princeton University Press.

CHAPTER 1. FROM DROUGHT TO DELUGE

Carle, David. 2004. *Introduction to Water in California*. Berkeley: University of California Press.

Gatewood, J. S. 1962. The meteorologic phenomenon of drought in the Southwest. U.S. Geological Survey, Professional Paper 372-A.

Gatewood, J. S., A. Wilson, H. E. Thomas, and L. R. Kister. 1964. General effects of drought on water resources of the Southwest, 1942–1956. U.S. Geological Survey, Professional Paper 372-B.

Powell, J. L. 2008. *Dead Pool: Lake Powell, Global Warming, and the Future of Water in the West*. Berkeley: University of California Press.

Schoner, T., and S. E. Nicholson. 1989. The relationship between California rainfall and ENSO events. *Journal of Climate* 2 (11): 1258–69.

Schulman, E. 1937. Selection of trees for climatic study. *Tree Ring Bulletin* 3 (3): 22–23.

Steinbeck, J. 1952. *East of Eden*. New York: Penguin Press.

Straka, T. 2008. Biological portrait, Edmund P. Schulman. *Forest History Today* (Spring): 46–49.

Trenberth, K. E. 1997. The definition of El Niño. *Bulletin of the American Meteorological Society* 78: 2271–77.

CHAPTER 2. THE 1861–1862 FLOODS

Brewer, W. H. (1930) 1966. *Up and Down California in 1860–1864*. Edited by F. P. Farquahar. Reprint, Berkeley: University of California Press.
Cleland, R. G., and J. Brooks, eds. 1955. *A Mormon Chronicle: The Diaries of John D. Lee, 1848–1876*. San Marino, Calif.: Huntington Library Press.
Dettinger, M. D., F. M. Ralph, T. Das, P. J. Neiman, and D. Cayan. 2011. Atmospheric rivers, floods, and the water resources of California. *Water* 3 (2): 445–78.
Engstrom, W. N. 1996. The California storm of January 1862. *Quaternary Research* 46: 141–48.
"The Floods of 1861–1862 in Northern California." 1956. *Oakland Tribune,* Jan. 8.
Gilbert, G. K. 1917. Hydraulic-mining debris in the Sierra Nevada. U.S. Geological Survey, Professional Paper 105.
Jaffe, B. E., R. E. Smith, and L. Z. Torresan. 1998. Sedimentation and bathymetric change in San Pablo Bay: 1856–1983. U.S. Geological Survey, Open-File Report 98–759.
Lynch, H. B. 1931. Rainfall and stream run-off in Southern California since 1769. Metropolitan Water District of Southern California, Los Angeles.
Mock, C. J. 2006. Some comments on reconstructing the historical climate of California, USA. *Societa Italiana de Fisica* 29 (1): 55–63.
Sidler, W. A. 1968. "Agua Mansa and the Flood of January 22, 1862, Santa Ana River." San Bernardino County Flood Control District.
Swetnam, T. W., C. H. Baisan, A. C. Caprio, P. M. Brown, R. Touchan, R. S. Anderson, and D. J. Hallett. 2009. Multi-millennial fire history of the Giant Forest, Sequoia National Park, California. *Fire Ecology* 5 (3): 120–50.
Van Tilburg Clark, W. 1973. *The Journals of Alfred Doten, Volume One, 1849–1903*. Reno: University of Nevada Press.
Ventura County Historical Society. 1958. The winter of 1861–1862. *Quarterly* (Nov.): 14–19.

CHAPTER 3. THE GREAT DROUGHTS OF
THE TWENTIETH CENTURY

Cayan, D. R., K. T. Redmond, and L. G. Riddle. 1999. ENSO and hydrologic extremes in the western United States. *Journal of Climate* 12: 2881–93.
Cook, B. I., R. L. Miller, and R. Seager. 2008. Dust and sea surface temperature forcing of the 1930s "Dust Bowl" drought. *Geophysical Research Letters* 35: L08710. doi:10.1029/2008GL033486.

Egan, T. 2006. *The Worst Hard Time: The Untold Story of Those Who Survived the Great American Dust Bowl.* Boston: Houghton Mifflin.

Harding, S. 1965. Recent variations in the water supply of the western Great Basin. Berkeley, University of California Water Resources Center, Archives Series Report 16.

Schubert, S. D., M. J. Suarez, P. J. Pegion, R. D. Koster, and J. T. Bacmeister. 2004. On the cause of the 1930s Dust Bowl. *Science,* Mar. 19.

CHAPTER 4. WHY IS CLIMATE SO VARIABLE
IN THE WEST?

Bjerknes, J. 1969. Atmospheric teleconnections from the equatorial Pacific. *Monthly Weather Review* 97: 163–72.

Cayan, D. R., M. D. Dettinger, H. F. Diaz, and N. E. Graham. 1998. Decadal variability of precipitation over western North America. *Journal of Climate* 11: 3148–66.

Cayan, D. R., and D. H. Peterson. 1989. The influence of North Pacific atmospheric circulation on streamflow in the West. In *Aspects of Climate Variability in the Pacific and the Western Americas,* edited by D. H. Peterson, 375–98. Washington, D.C.: American Geophysical Union.

Cayan, D. R., K. T. Redmond, and L. G. Riddle. 1999. ENSO and hydrologic extremes in the western United States. *Journal of Climate* 12 (9): 2881–93.

Dettinger, M. 2004. Fifty-two years of "pineapple-express" storms across the west coast of North America. U.S. Geological Survey, Scripps Institution of Oceanography for the California Energy Commission, PIER Energy-Related Environmental Research, CEC-500-2005-004.

Dettinger, M. D., and D. R. Cayan. 2003. Interseasonal covariability of Sierra Nevada streamflow and San Francisco Bay salinity. *Journal of Hydrology* 277 (3–4): 164.

Dettinger, M. D., F. M. Ralph, T. Das, P. J. Neiman, and D. Cayan. 2011. Atmospheric rivers, floods, and the water resources of California. *Water* 3 (2): 445–78.

Fagan, B. 1999. *Floods, Famines, and Emperors: El Niño and the Fate of Civilizations.* New York: Basic Books.

Francis, R. C., and S. R. Hare. 1994. Decadal-scale regime shifts in the large marine ecosystems of the northeast Pacific: A case for historical science. *Fisheries Oceanography* 3: 279–91.

Glanz, M. H. 2001. *Currents of Change: El Niño's Impact on Climate and Society.* 2nd edition. Cambridge, Engl.: Cambridge University Press.

Hare, S. R., N. J. Mantua, and R. C. Francis. 1999. Inverse production regimes: Alaskan and west coast salmon. *Fisheries* 24 (1): 6–14.

Latif, M., and T. P. Barnett. 1994. Causes of decadal climate variability over the North Pacific and North America. *Science* 266: 634–37.

Mantua, N.J., and S.R. Hare. 2002. The Pacific decadal oscillation. *Journal of Oceanography* 58: 35–42.

Mantua, N.J., S.R. Hare, Y. Zhang, J.M. Wallace, and R.C. Francis. 1997. A Pacific interdecadal climate oscillation with impacts on salmon production. *Bulletin of the American Meteorological Society* 78: 1069–79.

Muir, J. 1894. *The Mountains of California.* New York: The Century Co. www .sierraclub.org/john_muir_exhibit/writings/the_mountains_of_california.

Powell, J.L. 2008. *Dead Pool: Lake Powell, Global Warming, and the Future of Water in the West.* Berkeley: University of California Press.

Redmond, K.T., and R.W. Koch. 1991. Surface climate and streamflow variability in the western United States and their relationship to large-scale circulation indices. *Water Resources Research* 27: 2381–99.

Schoner, T., and S.E. Nicholson. 1989. The relationship between California rainfall and ENSO events. *Journal of Climate* 2 (11): 1258–69.

Walker, Sir Gilbert. 1923. Correlations in seasonal variations of weather. *India Meteorological Service Memoirs* 24 (4): 22.

PART II. A CLIMATE HISTORY OF THE AMERICAN WEST

Fagan, B. 2004. *The Long Summer: How Climate Changed Civilization.* New York: Basic Books.

CHAPTER 5. READING THE PAST

Bradley, R.S. 1999. *Paleoclimatology: Reconstructing Climates of the Quaternary.* 2nd edition. Waltham, Mass.: Academic Press.

Feynman, R. 1965. *The Character of Physical Law.* Cambridge, Mass.: MIT Press.

Hughes, M.K. 2002. Dendrochronology in climatology: The state of the art. *Dendrochronologia* 20 (1–2): 95–116.

Hughes, M.K., and L.J. Graumlich. 1996. Climatic variations and forcing mechanisms of the last 2000 years. In *Multi-millennial Dendro-climatic Studies from the Western United States,* 109–24. Berlin: Springer.

Imbrie, J., and K.P. Imbrie. 1986. *Ice Ages: Solving the Mystery.* Cambridge, Mass.: Harvard University Press.

Ingram, B.L., M.E. Conrad, and J.C. Ingle. 1996. Stable isotope and salinity systematics in estuarine waters and carbonates: San Francisco Bay. *Geochimica et Cosmochimica Acta* 60: 455–67.

Ingram, B.L., and J.R. Southon. 1996. Reservoir ages in eastern Pacific coastal and estuarine waters. *Radiocarbon* 38: 573–82.

LaMarche, V.C. 1974. Paleoclimatic inferences from long tree-ring records. *Science* 183: 1043–48.

Lamb, H. H. 1977. *Climate: Present, Past and Future.* London: Methuen.

Long, C. J., C. Whitlock, P. J. Bartlein, and S. H. Millspaughi. 1998. A 9000-year fire history from the Oregon coast range, based on a high-resolution charcoal study. *Canadian Journal of Forest Research* 28 (5): 774–87.

Macdougall, D. 2004. *Frozen Earth: The Once and Future Story of Ice Ages.* Berkeley: University of California Press.

———. 2009. *Nature's Clocks: How Scientists Measure the Age of Almost Everything.* Berkeley: University of California Press.

Ruddiman, W. F. 2007. *Earth's Climate: Past and Future.* New York: W. H. Freeman.

Schulman, E. 1937. Selection of trees for climatic study. *Tree Ring Bulletin* 3 (3): 22–23.

———. 1945. Tree-ring hydrology of the Colorado Basin. *University of Arizona Bulletin* 16 (4): 1–51.

CHAPTER 6. FROM ICE TO FIRE

Barron, J. A., and L. Anderson. 2011. Enhanced late Holocene ENSO/PDO expression along the margins of the eastern North Pacific. *Quaternary International* 235: 3–12.

Barron, J. A., L. Heusser, T. Herbert, and M. Lyle. 2003. High-resolution climatic evolution of coastal northern California during the past 16,000 years. *Paleoceanography* 18: 1020–34.

Benson, L. V. 1981. Paleoclimatic significance of lake-level fluctuations in the Lahontan Basin. *Quaternary Research* 16: 390–403.

Benson, L. V., D. R. Currey, R. I. Dorn, K. R. Lajoie, C. G. Oviatt, S. W. Robinson, G. I. Smith, and S. Stine. 1990. Chronology of expansion and contraction of four Great Basin lake systems during the past 35,000 years. In a special issue, "Paleolakes and Paleo-oceans," edited by P. A. Meyers and L. V. Benson. *Palaeogeography, Palaeoclimatology, Palaeoecology* 78: 241–86.

Digital Atlas of Idaho. Managed by Idaho State University Department of Geosciences and hosted by the Idaho Museum of Natural History, Pocatello, ID. imnh .isu.edu/digitalatlas.

Edlund, E. G. 1996. Late quaternary environmental history of montane forests of the Sierra Nevada, California. Ph.D. diss., Geography Department, University of California, Berkeley.

Edlund, E. G., and A. R. Byrne. 1991. Climate, fire and late quaternary vegetation change in the central Sierra Nevada. In *Fire and the Environment: Ecological and Cultural Perspectives,* edited by S. C. Nodvin and T. A. Waldrop, 390–96. U.S.D.A. Forest Service General Technical Report SE-69.

Erlandson, J. M., D. J. Kennett, B. L. Ingram, D. A. Guthrie, D. P. Morris, M. A. Tveskov, G. J. West, and P. L. Walker. 1996. An archaeological and paleontological chronology for Daisy Cave (CA-SMI-261), San Miguel Island, California. *Radiocarbon* 38: 355–73.

Fagan, B. 1999. *Floods, Famines, and Emperors: El Niño and the Fate of Civilizations.* New York: Basic Books.

———. 2003. *Before California: An Archaeologist Looks at Our Earliest Inhabitants.* Lanham, Md.: Rowman and Littlefield.

Gilbert, G. K. 1890. Lake Bonneville. U.S. Geological Survey, Monograph 1.

Imbrie, J., and K. P. Imbrie. 1986. *Ice Ages: Solving the Mystery.* Cambridge, Mass.: Harvard University Press.

Ingram, B. L., and J. P. Kennett. 1995. Radiocarbon chronology and planktonic-benthic foraminiferal ^{14}C age differences in Santa Barbara Basin sediments, Hole 893A. *Proceedings of the Ocean Drilling Program* 146 (2): 19–30.

Kennett, D. J. 2005. *The Island Chumash: Behavioral Ecology of a Maritime Society.* Berkeley: University of California Press.

Kennett, D. J., and J. P. Kennett. 2000. Competitive and cooperative responses to climatic instability in coastal southern California. *American Antiquity* 65 (2): 379–95.

Kennett, D. J., J. P. Kennett, J. M. Erlandson, and K. G. Cannarioto. 2007. Human responses to Middle Holocene climate change on California's Channel Islands. *Quaternary Science Reviews* 26: 351–67.

Kennett, J. P., and B. L. Ingram. 1995. Paleoclimatic evolution of Santa Barbara Basin during the last 20 k.y.: Marine evidence from Hole 893A. *Proceedings of the Ocean Drilling Program* 146 (2): 309–28.

———. 1995. A 20,000-year record of ocean circulation and climate change from the Santa Barbara Basin. *Nature* 377: 510–14.

Lightfoot, K. G., and O. Parrish. 2009. *California Indians and Their Environment: An Introduction.* Berkeley: University of California Press.

Macdougall, D. 2004. *Frozen Earth: The Once and Future Story of Ice Ages.* Berkeley: University of California Press.

Viau, A. E., K. Gajewski, P. Fines, D. E. Atkinson, and M. C. Sawada. 2002. Widespread evidence of 1500 yr climate variability in North America during the past 14,000 yr. *Geology* 30: 455–58.

CHAPTER 7. THE "LONG DROUGHT"
OF THE MID-HOLOCENE

Antevs, E. 1938. *Rainfall and Tree Growth in the Great Basin.* Washington, D.C.: Carnegie Institution of Washington/American Geographical Society.

———. 1939. Precipitation and water supply in the Sierra Nevada, California. *Bulletin of the American Meteorological Society* 20: 89–91.

Atwater, B. F., C. W. Hedel, and E. J. Helley. 1977. Late quaternary depositional history, Holocene sea-level changes, and vertical crustal movement, southern San Francisco Bay, California. U.S. Geological Survey, Professional Paper 1014.

Barron, J. A., L. Heusser, T. Herbert, and M. Lyle. 2003. High-resolution climatic evolution of coastal northern California during the past 16,000 years. *Paleoceanography* 18 (1). doi:10.1029/2002PA000768.

Benson, L.V., and J.L. Bischoff. 1994. Isotope geochemistry of Owens Lake Core OL92. In *Core OL-92 from Owens Lake, Southeast California,* edited by G.I. Smith and J.L. Bischoff. U.S. Geological Survey, Open-File Report 93–683.

Benson, L.V., J.W. Burdett, M. Kashgarian, S.P. Lund, F.M. Phillips, and R.O. Rye. 1996. Climatic and hydrologic oscillations in the Owens Lake basin and adjacent Sierra Nevada, California. *Science* 274: 746–49.

Benson, L.V., M. Kashgarian, R. Rye, S. Lund, F. Paillet, J. Smoot, C. Kester, S. Mensing, D. Meko, and S. Lindstrom. 2002. Holocene multidecadal and multicentennial droughts affecting northern California and Nevada. *Quaternary Science Reviews* 21: 659–82.

Benson, L.V., and R.S. Thompson. 1987. Lake-level variation in the Lahontan basin for the past 50,000 years. *Quaternary Research* 28: 69–85.

Ely, L.L., Y. Enzel, V.R. Baker, and D.R. Cayan. 1993. A 5000-yr record of extreme floods and climate change in the southwestern United States. *Science* 262: 410–12.

Fagan, B. 1999. *Floods, Famines, and Emperors: El Niño and the Fate of Civilizations.* New York: Basic Books.

Fladmark, K.R. 1975. *A Paleoecological Model for Northwest Coast Prehistory.* Ottawa: National Museums of Canada.

Harding, S. 1965. Recent variations in the water supply of the western Great Basin. Berkeley: University of California Water Resources Center, Archives Series Report 16.

Kennett, D.J., B.J. Culleton, J.P. Kennett, J.M. Erlandson, and K.G. Cannariato. 2007. Middle Holocene climate change and human population dispersal in western North America. In *Climate Change and Cultural Dynamics: A Global Perspective on Mid-Holocene Transitions,* edited by D.G. Anderson, K.A. Maasch, and D.H. Sandweiss, 531–58. Waltham, Mass.: Academic Press/ Elsevier.

Lightfoot, K. 1997. Cultural construction of coastal landscapes: A middle Holocene perspective from San Francisco Bay. In *Archaeology of the California Coast during the Middle Holocene,* edited by J. Erlandson and M. Glassow, 129–41. Los Angeles: Cotsen Institute of Archaeology.

Lindstrom, S. 1990. Submerged tree stumps as indicators of mid-Holocene aridity in the Lake Tahoe basin. *Journal of California and Great Basin Anthropology* 12 (2): 146–57.

Moss, M.L., D.M. Peteet, and C. Whitlock. 2007. Mid-Holocene culture and climate on the northwest coast of North America. In *Climate Change and Cultural Dynamics: A Global Perspective on Mid-Holocene Transitions,* edited by D.G. Anderson, K.A. Maasch, and D.H. Sandweiss, 491–530. Waltham, Mass.: Academic Press/Elsevier.

Stine, S.W. 1990. Late Holocene fluctuations of Mono Lake, eastern California. *Palaeogeography, Palaeoclimatology, and Palaeoecology* 78: 333–81.

Anderson, R. S., and S. J. Smith. 1997. The sedimentary record of fire in Montane Meadows, Sierra Nevada, California, USA. In *Sediment Records of Biomass Burning and Global Change,* edited by J. Clarke, H. Cachier, J. G. Goldammer, and B. Stocks, 313–28. Berlin: Springer.

Gundmundsson, H. J. 1997. A review of the Holocene environmental history of Iceland. *Quaternary Science Reviews* 16: 81–92.

Holzhauser, H., M. Magny, and H. J. Zumbuhl. 2005. Glacier and lake level variations in west-central Europe over the last 3500 years. *The Holocene* 15: 789–801.

Hughes, M. K., and G. Funkhouser. 1998. Extremes of moisture availability reconstructed from tree-rings from recent millennia in the Great Basin of western North America. In *Impacts of Climate Variability on Forests,* edited by M. Beniston and J. L. Innes, 99–107. Berlin: Springer.

Ingram, B. L. 1998. Differences in radiocarbon age between shell and charcoal from a Holocene shellmound in northern California. *Quaternary Research* 49: 102–10.

Ingram, B. L., M. E. Conrad, and J. C. Ingle. 1996a. Stable isotope record of late Holocene salinity and river discharge in San Francisco Bay, California. *Earth and Planetary Science Letters* 141: 237–47.

———. 1996b. A 2000-yr record of Sacramento–San Joaquin River inflow to San Francisco Bay estuary, California. *Geology* 24: 331–34.

Ingram, B. L., and D. DePaolo. 1993. A 4300 yr strontium isotope record of estuarine paleosalinity in San Francisco Bay, CA. *Earth and Planetary Science Letters* 119: 103–19.

Ingram, B. L., and D. Sloan. 1992. Strontium isotopic composition of estuarine sediments as paleosalinity-paleoclimate indicator. *Science* 255: 68–72.

Jones, T. L., G. M. Brown, L. M. Raab, J. L. McVickar, W. G. Spaulding, D. J. Kennett, A. York, and P. L. Walker. 1999. Environmental imperatives reconsidered. *Current Anthropology* 40: 137–70.

Kennett, D. J., and J. P. Kennett. 2000. Competitive and cooperative responses to climatic instability in coastal southern California. *American Antiquity* 65 (2): 379–95.

LaMarche, V. C. 1974. Paleoclimatic inferences from long tree-ring records. *Science* 183: 1043–48.

Lamb, H. H. 1977. *Climate: Present, Past and Future.* London: Methuen.

Lightfoot, K. G., and E. M. Luby. 2002. Late Holocene in the greater San Francisco Bay Area: Temporal trends in the use and abandonment of shell mounds in the East Bay. In *Catalysts to Complexity: The Late Holocene on the California Coast,* edited by J. Erlandson and T. Jones, 263–81. Los Angeles: Cotsen Institute of Archaeology.

Long, C. J., C. Whitlock, P. J. Bartlein, and S. H. Millspaughi. 1998. A 9000-year fire history from the Oregon coast range, based on a high-resolution charcoal study. *Canadian Journal of Forest Research* 28 (5): 774–87.

MacDonald, G. M., A. A. Velichko, C. V. Kremenetski, O. K. Borisova, A. A. Goleva, A. A. Andreev, L. C. Cwynar, R. T. Riding, S. L. Forman, T. W. D. Edwards, R. Aravena, D. Hammarlund, J. M. Szeicz, and V. N. Gattaulin. 2000. Holocene treeline history and climate change across northern Eurasia. *Quaternary Research* 53: 302–11.

Matthes, F. 1939. Report of Committee on Glaciers. *Transactions of the American Geophysical Union* 20: 518–23.

Matthews, J. A., and P. Q. Dresser. 2008. Holocene variation glacier chronology of the Smortstabbtindam massif, Jotunheimen, southern Norway, and the recognition of century to millennial scale European neoglacial events. *The Holocene* 18: 181–201.

Mehringer, P. J., and P. E. Wigand. 1990. Comparison of late Holocene environments from woodrat middens and pollen: Diamond Craters, Oregon. In *Fossil Packrat Middens: The Last 40,000 Years of Biotic Changes,* edited by J. L. Betancourt, T. R. Van Devender, and P. S. Martin. Tucson: University of Arizona Press.

Mohr, J. A., C. Whitlock, and C. N. Skinner. 2000. Postglacial vegetation and fire history, eastern Klamath mountains, California, USA. *The Holocene* 10: 587–601.

Nederbragt, A. J., J. W. Thurow, and P. R. Brown. 2008. Paleoproductivity, ventilation, and organic carbon burial in the Santa Barbara Basin (ODP Site 893, off California) since the last glacial. *Paleoceanography* 23: PA1211. doi:10.1029/207PA001505.

Nelson, N. C. 1909. Shellmounds of the San Francisco Bay region. *University of California Publications in American Archaeology and Ethnology* 7: 309–56.

Schweikhardt, P., B. L. Ingram, K. G. Lightfoot, and E. M. Luby. 2011. Geochemical methods for inferring seasonal occupation of an estuarine shellmound: A case study from San Francisco Bay. *Journal of Archaeological Science* 38: 2301–12.

Starratt, S. W. 2004. Diatoms as indicators of late Holocene freshwater flow variations in the San Francisco Bay estuary, central California, USA. In *Proceedings of the Seventeenth International Diatom Symposium, Ottawa, Canada, 25th–31st August 2002,* edited by M. Poulin, 371–97. Bristol, U.K.: Biopress Limited.

Stine, S. W. 1990. Late Holocene fluctuations of Mono Lake, eastern California. *Palaeogeography, Palaeoclimatology and Palaeoecology* 78: 333–81.

Stockton, C. W., and D. M. Meko. 1975. A long-term history of drought occurrence in western United States as inferred from tree-rings. *Weatherwise* 28: 244–49.

Stott, L. D., K. G. Cannariato, R. Thunell, G. H. Haug, A. Koutavas, and S. Lund. 2004. Decline of surface temperature and salinity in the western tropical Pacific Ocean in the Holocene epoch. *Nature* 431: 56–59.

Wigand, P. E. 1997. A late Holocene pollen record from Lower Pahranagat Lake, southern Nevada, USA: High resolution paleoclimatic records and analysis of environmental responses to climate change. In *Proceedings of the Thirteenth Annual Pacific Climate (PACLIM) Workshop, Asilomar, California, April 14–17, 1996.* California Department of Water Resources Technical Report 53 of the Interagency Ecological Program for the Sacramento–San Joaquin Estuary.

Wigand, P.E., and D. Rhode. 2002. Great Basin vegetation history and aquatic systems: The last 150,000 years. In *Great Basin Aquatic Systems History,* edited by R. Hershler, D.B. Madsen, and D.R. Currey, 309–67. Washington, D.C.: Smithsonian Institution Press.

CHAPTER 9. THE GREAT "MEDIEVAL DROUGHT"

Adams, K.D. 2003. Age and paleoclimatic significance of late Holocene lakes in the Carson Sink, NV, USA. *Quaternary Research* 60: 294–306.

Benson, L., K. Petersen, and J. Stein. 2007. Anasazi (pre-Columbian Native-American) migrations during the middle-12th and late 13th centuries—Were they drought induced? *Climatic Change* 83: 187–213.

Bradley, R.S. 2000. 1000 years of climate changes. *Science* 288: 1353–54.

Bradley, R.S., M.K. Hughes, and H.F. Diaz. 2003. Climate in Medieval time. *Science* 302: 404–5.

Byrne, R., B.L. Ingram, S. Starratt, F. Malamud-Roam, J.N. Collins, and M.E. Conrad. 2001. Carbon isotope, diatom, and pollen evidence for late Holocene salinity change in a brackish marsh in the San Francisco estuary. *Quaternary Research* 55: 66–76.

Cook, E.R., D.M. Meko, D.W. Stahle, and M.K. Cleaveland. 1999. Drought reconstructions for the continental United States. *Journal of Climate* 12 (4): 1145–62.

Cook, E.R., C. Woodhouse, C.M. Eakin, D.M. Meko, and D.W. Stahle. 2004. Long-term aridity changes in the western United States. *Science* 306: 1015–18.

Earle, C.J. 1993. Asynchronous droughts in California streamflow as reconstructed from tree rings. *Quaternary Research* 39: 290–99.

Fagan, B. 2008. *The Great Warming: Climate Change and the Rise and Fall of Civilizations.* New York: Bloomsbury Press.

Graumlich, L.J. 1993. A 1000-year record of temperature and precipitation in the Sierra Nevada. *Quaternary Research* 39: 249–55.

Grayson, D.K. 1993. *The Desert's Past: A Natural Prehistory of the Great Basin.* Washington, D.C.: Smithsonian Institution Press.

Haskins, C.H. 1927. *The Renaissance of the Twelfth Century.* Cambridge, Mass.: Harvard University Press.

Hodell, D.A., J.H. Curtis, and M. Brenner. 1995. A possible role of climate in the collapse of the classic Maya civilization. *Nature* 375: 391–94.

Hughes, M.K., and H.F. Diaz. 1994. Was there a "Medieval warm period" and if so, where and when? *Climate Change* 26: 109–42.

Ingram, B.L., M.E. Conrad, and J.C. Ingle. 1996a. Stable isotope record of late Holocene salinity and river discharge in San Francisco Bay, California. *Earth and Planetary Science Letters* 141: 237–47.

———. 1996b. A 2000-yr record of Sacramento–San Joaquin River inflow to San Francisco Bay estuary, California. *Geology* 24: 331–34.

Johnson, G. 1996. The Anasazi "collapse." *New York Times,* Aug. 20, p. 1.

Lamb, H. H. 1965. The early Medieval warm epoch and its sequel. *Palaeogeography, Palaeoclimatology, and Palaeoecology* 1: 13–37.

Lambert, P. 1993. Health in prehistoric populations of the Santa Barbara Channel Islands. *American Antiquity* 68: 509–22.

Leavitt, S. W. 1994. Major wet interval in White Mountains Medieval warm period evidenced in ^{13}C of Bristlecone Pine tree rings. *Climatic Change* 26: 299–307.

Lightfoot, K. G., and E. M. Luby. 2002. Late Holocene in the greater San Francisco Bay Area: Temporal trends in the use and abandonment of shell mounds in the East Bay. In *Catalysts to Complexity: The Late Holocene on the California Coast,* edited by J. Erlandson and T. Jones, 263–81. Los Angeles: Cotsen Institute of Archaeology.

Malamud-Roam, F. 2002. A late Holocene history of vegetation change in San Francisco estuary marshes using stable carbon isotopes and pollen analysis. Ph.D. diss., University of California, Berkeley.

Malamud-Roam, F., and B. L. Ingram. 2001. Carbon isotopic compositions of plants and sediments of tide marshes in the San Francisco estuary. *Journal of Coastal Research* 17 (1): 17–19.

———. 2004. Late Holocene d^{13}C and pollen records of paleosalinity from tidal marshes in the San Francisco estuary. *Quaternary Research* 62: 134–45.

Meko, D. M., M. D. Therrell, C. H. Baisan, and M. K. Hughes. 2001. Sacramento River flow reconstructed to A.D. 869 from tree rings. *Journal of American Water Research Association* 37: 1029–39.

Meko, D. M., C. A. Woodhouse, C. H. Baisan, T. Knight, J. J. Lukas, M. K. Hughes, and M. W. Salzer. 2007. Medieval drought in the upper Colorado River basin. *Geophysical Research Letters* 34: L10705. doi:10.1029/2007GL029988.

Mensing, S. A., L. V. Benson, M. Kashgarian, and S. Lund. 2004. A Holocene pollen record of persistent droughts from Pyramid Lake, Nevada, USA. *Quaternary Research* 62: 29–38.

Millar, C. I., J. C. King, R. D. Westfall, H. A. Alden, and D. L. Delany. 2006. Late Holocene forest dynamics, volcanism, and climate change at Whitewing Mountain and San Joaquin Ridge, Mono County, Sierra Nevada, CA, USA. *Quaternary Research* 66: 273–87.

Olson, J. E., and E. G. Bourne, eds. 1906. *The Northmen, Columbus and Cabot, 985–1503: The Voyages of the Northmen; The Voyages of Columbus and of John Cabot.* New York: Charles Scribner's Sons.

Powell, J. L. 2008. *Dead Pool: Lake Powell, Global Warming, and the Future of Water in the West.* Berkeley: University of California Press.

Raab, I. M., and D. O. Larson. 1997. Medieval climatic anomaly and punctuated cultural evolution in coastal Southern California. *American Antiquity* 62: 319–36.

Schweikhardt, P., D. Sloan, and B. L. Ingram. 2008. Early Holocene evolution of San Francisco estuary, Northern California. *Journal of Coastal Research* 26: 704–13.

Starratt, S. W. 2004. Diatoms as indicators of late Holocene freshwater flow variations in the San Francisco Bay estuary, central California, USA. In *Proceedings of the Seventeenth International Diatom Symposium, Ottawa, Canada, 25th–31st August 2002*, edited by M. Poulin, 371–97. Bristol, U.K.: Biopress Limited.

Stine, S. 1990. Late Holocene fluctuations of Mono Lake, eastern California. *Palaeogeography, Palaeoclimatology and Palaeoecology* 78: 333–81.

———. 1994. Extreme and persistent drought in California and Patagonia during Medieval time. *Nature* 369: 546–49.

Stockton, C. W., and D. M. Meko. 1975. A long-term history of drought occurrence in western United States as inferred from tree-rings. *Weatherwise* 28: 244–49.

Swetnam, T. W. 1993. Fire history and climate change in Giant Sequoia groves. *Science* 262: 885–89.

Swetnam, T. W., C. H. Baisan, A. C. Caprio, P. M. Brown, R. Touchan, R. S. Anderson, and D. J. Hallett. 2009. Multi-millennial fire history of the Giant Forest, Sequoia National Park, California. *Fire Ecology* 5 (3): 120–50.

Walker, P. 1989. Cranial injuries as evidence of violence in prehistoric California. *American Journal of Physical Anthropology* 80: 51–61.

Woodhouse, C., D. M. Meko, G. M. MacDonald, D. W. Stahle, and E. R. Cook. 2010. A 1,200-year perspective of 21st century drought in southwestern North America. *Proceedings of the National Academy of Science* 107: 21283–88.

CHAPTER 10. THE LITTLE ICE AGE

Barron, J. A., D. Bukry, and D. Field. 2008. Santa Barbara Basin diatom and silicoflagellate records suggest coincidence of cooler SST with widespread occurrence of drought in the West during the past 2200 years. *Quaternary International* 215: 34–44. doi:10.1016/j.quaint.2008.08.007.

Benson, L., M. Kashgarian, R. Dye, S. Lund, F. Paillet, J. Smooth, C. Kester, S. Mensing, D. Meko, and S. Lindström. 2002. Holocene multidecadal and multicentennial droughts affecting Northern California and Nevada. *Quaternary Science Reviews* 21: 659–82.

Biondi, F., C. Isaacs, M. K. Hughes, D. R. Cayan, and W. H. Berger. 2000. The near-1600 dry/wet knockout: Linking terrestrial and near-shore ecosystems. In *Proceedings of the Twenty-Fourth Annual Climate Diagnostics and Prediction Workshop*. U.S. Department of Commerce, National Oceanographic and Atmospheric Administration.

Boxt, M. A., L. M. Raab, O. K. Davis, and K. O. Pope. 1999. Extreme late Holocene climate change in coastal California. *Pacific Coast Archeological Society Quarterly* 35: 25–37.

Byrne, R., B. L. Ingram, S. Starratt, F. Malamud-Roam, J. N. Collins, and M. E. Conrad. 2001. Carbon isotope, diatom, and pollen evidence for late Holocene salinity change in a brackish marsh in the San Francisco estuary. *Quaternary Research* 55: 66–76.

Daniels, M. L., R. S. Anderson, and C. Whitlock. 2005. Vegetation and fire history since the late Pleistocene for the Trinity Mountains, northwestern California, USA. *The Holocene* 15: 1062–71.

Davis, O. 1992. Rapid climate change in coastal Southern California inferred from pollen analysis of San Joaquin marsh. *Quaternary Research* 37: 89–100.

Earle, C.J., and H. C. Fritts. 1986. Reconstructing riverflow in the Sacramento basin since 1560. Tucson: University of Arizona Laboratory of Tree-Ring Research.

Edlund, E.G., and A. R. Byrne. 1991. Climate, fire and late Quaternary vegetation change in the central Sierra Nevada. In *Fire and the Environment: Ecological and Cultural Perspectives,* edited by S. C. Nodvin and T. A. Waldrop, 390–96. U.S.D.A. Forest Service General Technical Report SE-69.

Fagan, B. 2000. *The Little Ice Age: How Climate Made History, 1300–1850.* New York: Basic Books.

Graham, N. E. 2004. Late Holocene teleconnections between tropical Pacific climate variability and precipitation in the western USA: Evidence from proxy records. *The Holocene* 14: 436–47.

Graumlich, L. J. 1993. A 1000-year record of temperature and precipitation in the Sierra Nevada. *Quaternary Research* 39: 249–55.

Hughes, M. K., and G. Funkhouser. 1998. Extremes of moisture availability reconstructed from tree-rings from recent millennia in the Great Basin of western North America. In *Impacts of Climate Variability on Forests,* edited by M. Beniston and J. L. Innes, 99–107. Berlin: Springer.

Hughes, M. K., and L. J. Graumlich. 1996. Climatic variations and forcing mechanisms of the last 2000 years. In *Multi-millennial Dendro-climatic Studies from the Western United States,* 109–24. Berlin: Springer.

Ingram, B. L., M. E. Conrad, and J. C. Ingle. 1996a. Stable isotope record of late Holocene salinity and river discharge in San Francisco Bay, California. *Earth and Planetary Science Letters* 141: 237–47.

———. 1996b. A 2000-yr record of Sacramento–San Joaquin River inflow to San Francisco Bay estuary, California. *Geology* 24: 331–34.

Ingram, B. L., and D. DePaolo. 1993. A 4300 yr strontium isotope record of estuarine paleosalinity in San Francisco Bay, CA. *Earth and Planetary Science Letters* 119: 103–19.

Ingram, B. L., and D. Sloan. 1992. Strontium isotopic composition of estuarine sediments as paleosalinity-paleoclimate indicator. *Science* 255: 68–72.

Lamb, H. H. 1977. *Climate: Present, Past and Future.* London: Methuen.

Long, C. J., C. Whitlock, P. J. Bartlein, and S. H. Millspaughi. 1998. A 9000-year fire history from the Oregon coast range, based on a high-resolution charcoal study. *Canadian Journal of Forest Research* 28 (5): 774–87.

Meko, D. M., M. D. Therrell, C. H. Baisan, and M. K. Hughes. 2001. Sacramento River flow reconstructed to A.D. 869 from tree rings. *Journal of American Water Research Association* 37: 1029–39.

Mensing, S. A., R. Byrne, and J. Michaelsen. 1999. A 560-year record of Santa Ana fires reconstructed from charcoal deposited in the Santa Barbara Basin, California. *Quaternary Research* 51: 295–305.

Nederbragt, A. J., J. W. Thurow, and P. R. Brown. 2008. Paleoproductivity, ventilation, and organic carbon burial in the Santa Barbara Basin (ODP Site 893, off California) since the last glacial. *Paleoceanography* 23: PA1211. doi:10.1029/207PA001505.

Schimmelmann, A., C. B. Lange, and H.-C. Li. 2002. Major flood events of the past 2,000 years recorded in Santa Barbara Basin. In *Proceedings of the Eighteenth Annual Pacific Climate (PACLIM) Workshop*, edited by G. J. West and L. D. Buffaloe, 83–103. Technical Report 68 of the Interagency Ecological Program for the San Francisco Estuary, California Department of Water Resources.

Schimmelmann, A., C. B. Lange, and B. L. Meggers. 2003. Palaeoclimatic and archaeological evidence for a ~200-yr recurrence of floods and droughts linking California, Mesoamerica and South America over the past 2000 years. *The Holocene* 13 (5): 763–78.

Schimmelmann, A., M. Zhao, C. C. Harvey, and C. B. Lange. 1998. A large California flood and correlative global climatic events 400 years ago. *Quaternary Research* 49: 51–61.

Schonher, T., and S. E. Nicholson. 1989. The relationship between California rainfall and ENSO events. *Journal of Climate* 2: 1258–69.

Starratt, S. W. 2004. Diatoms as indicators of late Holocene freshwater flow variations in the San Francisco Bay estuary, central California, USA. In *Proceedings of the Seventeenth International Diatom Symposium, Ottawa, Canada, 25th–31st August 2002*, edited by M. Poulin, 371–97. Bristol, U.K.: Biopress Limited.

Stine, S. W. 1990. Late Holocene fluctuations of Mono Lake, eastern California. *Palaeogeography, Palaeoclimatology and Palaeoecology* 78: 333–81.

Sullivan, D. J. 1982. Prehistoric flooding in the Sacramento Valley: Stratigraphic evidence from Little Packer Lake, Glenn County, California. Master's thesis, University of California, Berkeley.

Woodhouse, C. A., S. T. Gray, and D. M. Meko. 2006. Updated streamflow reconstructions for the upper Colorado River basin. *Water Resources Research* 42: W05415. doi:10.1029/2005WR004455.

CHAPTER 11. WHY CLIMATE CHANGES

Alley, Richard. 2000. *The Two-Mile Time Machine: Ice Cores, Abrupt Climate Change, and Our Future*. Princeton, N.J.: Princeton University Press.

Clement, A. C., R. Seager, and M. A. Cane. 2000. Suppression of El Niño during the mid-Holocene by changes in the Earth's orbit. *Paleoceanography* 15: 731–37.

Cobb, K. M., C. D. Charles, H. Cheng, and R. L. Edwards. 2003. El Niño/Southern Oscillation and tropical Pacific climate during the last millennium. *Nature* 424: 271–76.

Croll, J. 1875. *Climate and Time in Their Geological Relations: A Theory of Secular Changes of the Earth's Climate.* London: Daldy, Isbister.

Davis, O. 1994. The correlation of summer precipitation in the southwestern U.S.A. with isotopic records of solar activity during the medieval warm period. *Earth and Environmental Science* 26: 271–87.

Debret, M., V. Bout-Roumazeilles, F. Grousset, M. Desmet, J. F. McManus, N. Massei, D. Sebag, J. R. Petit, Y. Copard, and A. Trentesaux. 2007. The origin of the 1500-year climate cycles in Holocene North-Atlantic records. *Climate of the Past* 3: 569–75.

Eddy, J. A. 1976. The Maunder minimum. *Science* 192: 1189–1202.

Emiliani, C. 1956. On paleotemperatures of Pacific bottom waters. *Science* 123: 460–61.

———. 1957. Temperature and age analysis of deep sea cores. *Science* 125: 383–85.

Fallon, S. J., M. T. McCulloch, and C. Alibert. 2003. Examining water temperature proxies in Porites corals from the Great Barrier Reef: A cross-shelf comparison. *Coral Reefs* 22 (4): 389–404.

Franklin, B. 1784. Meteorological imaginations and conjectures. *Memoirs of the Literary and Philosophical Society of Manchester* 2: 357–61.

Graham, N. E. 2004. Late Holocene teleconnections between tropical Pacific climate variability and precipitation in the western USA: Evidence from proxy records. *The Holocene* 14: 436–47.

Graham, N. E., M. K. Hughes, C. M. Ammann, K. M. Cobb, M. P. Hoerling, D. J. Kennett, J. P. Kennett, B. Rein, L. Stott, P. E. Wigand, and Taiyi Xu. 2007. Tropical Pacific midlatitude teleconnections in Medieval times. California Energy Commission, PIER Energy-Related Environmental Research Program, CEC-500-2007-022.

Hays, J. D., J. Imbrie, and N. J. Shackleton. 1976. Variations in the Earth's Orbit: Pacemaker of the Ice Ages. *Science* 194: 1121–32.

Hughes, M. K., and L. J. Graumlich. 1996. Climatic variations and forcing mechanisms of the last 2000 years. In *Multi-millennial Dendro-climatic Studies from the Western United States,* 109–24. Berlin: Springer.

Imbrie, J., and K. P. Imbrie. 1986. *Ice Ages: Solving the Mystery.* Cambridge, Mass.: Harvard University Press.

Macdougall, D. 2004. *Frozen Earth: The Once and Future Story of Ice Ages.* Berkeley: University of California Press.

Mann, M. E., M. A. Cane, S. E. Zebiak, and A. Clement. 2005. Volcanic and solar forcing of the tropical Pacific over the past 1000 years. *Journal of Climate* 18: 447–56.

Meehl, G. A., J. M. Arblaster, K. Matthes, F. Sassi, and H. van Loon. 2009. Amplifying the Pacific climate system response to a small 11-year solar cycle forcing. *Science* 325: 1114–18.

Milankovitch, M. (1941) 1969. *Canon of Insolation and the Ice-Age Problem.* Jerusalem: Israel Program for Scientific Translations.

Moy, C. M., G. O. Seltzer, D. T. Rodbell, and D. M. Anderson. 2002. Variability of El Niño/Southern Oscillation activity at millennial timescales during the Holocene epoch. *Nature* 420: 162–65.

Petit, J. R., J. Jourzel, D. Raynaud, N. I. Barkov, J. M. Barnola, I. Basile, M. Bender, J. Chappellaz, M. Davis, G. Delaygue, M. Delmotte, V. M. Kotlyakov, M. Legrand, V. Y. Lipenkov, C. Lorius, L. Pepin, C. Ritz, E. Saltzman, and M. Stievenard. 1999. Climate and atmospheric history of the past 420,000 years from the Vostok ice core, Antarctica. *Nature* 399: 429–36.

Quiroz, R. S. 1983. The isolation of stratospheric temperature change due to the El Chichon volcanic eruption from non-volcanic signals. *Journal of Geophysical Research* 88: 6773–80.

Rampino, M. R., and S. Self. 1992. Volcanic winter and accelerated glaciation following the Toba super-eruption. *Nature* 359: 50–52.

Reimer, P. J., M. G. L. Baillie, E. Bard, A. Bayliss, J. W. Beck, C. Bertrand, P. G. Blackwell, C. E. Buck, G. Burr, K. B. Cutler, P. E. Damon, R. L. Edwards, R. G. Fairbanks, M. Friedrich, T. P. Guilderson, K. A. Hughen, B. Kromer, F. G. McCormac, S. Manning, C. Bronk Ramsey, R. W. Reimer, S. Remmele, J. R. Southon, M. Stuiver, S. Talamo, F. W. Taylor, J. van der Plicht, C. E. Weyhenmeyer. 2004. IntCal04 terrestrial radiocarbon age calibration, 0–26 cal kyr BP. *Radiocarbon* 46: 1029–58.

Robock, A. 1979. The "Little Ice Age": Northern Hemisphere average observations and model calculations. *Science* 206: 1402–04.

———. 2000. Volcanic eruptions and climate. *Reviews of Geophysics* 38: 191–219.

Robock, A., and M. P. Free. 1996. The volcanic record in ice cores for the past 2000 years. In *Climatic Variations and Forcing Mechanisms of the Last 2000 Years,* edited by P. D. Jones, R. S. Bradley, and J. Jouzel, 533–46. New York: Springer-Verlag.

Rodbell, D. T., G. O. Seltzer, D. M. Anderson, M. B. Abbott, D. B. Enfield, and J. H. Newman. 1999. An ~15,000 year record of El Niño–driven alluviation in southwestern Ecuador. *Science* 283: 516–20.

Tudhope, A. W., C. P. Chilcott, M. T. McCulloch, E. R. Cook, J. Chappell, R. M. Ellam, D. W. Lea, J. M. Lough, and G. B. Shimmield. 2001. Variability in the El Niño–Southern Oscillation through a glacial-interglacial cycle. *Science* 291: 1511–17.

PART III. A GROWING WATER CRISIS

Philander, S. G. 1998. *Is the Temperature Rising? The Uncertain Science of Global Warming.* Princeton, N.J.: Princeton University Press.

CHAPTER 12. THE HYDRAULIC ERA

Brewer, W. H. 2003. *Up and Down California in 1860–1864.* Berkeley: University of California Press.

Bryan, K. 1923. Geology and groundwater resources of the Sacramento Valley, CA. U.S. Geological Survey, Water-Supply Paper 495.

California Department of Water Resources (DWR). 2003. California's groundwater. Bulletin 118.

———. 2005. California Water Plan: A Framework for Action. Bulletin 160–05.

Carle, D. 2003. *Water and the California Dream: Choices for the New Millennium.* San Francisco: Sierra Club Books.

Cohen, M. J., C. Henges-Jeck, and G. Castillo-Moreno. 2001. A preliminary water balance for the Colorado River delta, 1992–1998. *Journal of Arid Environments* 49: 35–48.

Dettman, D. L., K. W. Flessa, P. D. Roopnarine, B. R. Schöne, and D. H. Goodwin. 2004. The use of oxygen isotope variation in shells of estuarine mollusks as a quantitative record of seasonal and annual Colorado River discharge. *Geochimica et Cosmochimica Acta* 68: 1253–63.

Dominy, F. 1997. Interview. In *Cadillac Desert: Water and the Transformation of Nature. Program Three: The Mercy of Nature.* PBS documentary directed by Jon Else.

Fagan, B. 2003. *Before California: An Archaeologist Looks at Our Earliest Inhabitants.* Lanham, Md.: Rowman and Littlefield Publishers.

Filippone, C., and S. A. Leake. 2005. Time scales in the sustainable management of water resources. *Southwest Hydrology* 4 (1): 16–17.

Finney, B. P., I. Gregory-Eaves, M. S. V. Douglas, and J. P. Smol. 2002. Fisheries productivity in the northeastern Pacific Ocean over the past 2,200 years. *Nature* 416: 729–33.

Gilbert, G. K. 1917. Hydraulic mining in the Sierra Nevada. U.S. Geological Survey, Professional Paper 105.

Hanak, E., J. Lund, A. Dinar, B. Gray, R. Howitt, J. Mount, P. Moyle, and B. Thompson. 2011. *Managing California's Water: From Conflict to Reconciliation.* San Francisco: Public Policy Institute of California.

Haury, E. W. 1978. *The Hohokam: Desert Farmers and Craftsmen.* Tucson: University of Arizona Press.

Hundley, N., Jr. 2001. *The Great Thirst: Californians and Water, a History.* Rev. ed. Berkeley: University of California Press.

Jaffe, B. E., R. E. Smith, and L. Z. Torresan. 1998. Sedimentation and bathymetric change in San Pablo Bay: 1856–1983. U.S. Geological Survey, Open-File Report 98–759.

Jassby, A. D., W. J. Kimmerer, S. G. Monismith, C. Armor, J. E. Cloern, T. M. Powell, J. R. Schubel, and T. J. Vendlinski. 1995. Isohaline position as a habitat indicator for estuarine populations. *Ecological Applications* 5 (1): 272–89.

Knowles, N. 2000. Modelling the hydroclimate of the San Francisco Bay–Delta estuary and watershed. Ph.D. diss., University of California, San Diego.

Leopold, A. 1970. *A Sand County Almanac: With Essays on Conservation from Round River.* New York: Ballantine Books.

Lichatowich, J. 1999. *Salmon without Rivers: A History of the Pacific Salmon Crisis.* Washington, D.C.: Island Press.

Lightfoot, K. G., and O. Parrish. 2009. *California Indians and Their Environment: An Introduction.* Berkeley: University of California Press.

Luecke, D. F., J. Pitt, C. Congdon, E. Glenn, C. Valdes-Casillas, and M. Briggs. 1999. *A Delta Once More: Restoring Riparian and Wetland Habitat in the Colorado River Delta.* Washington, D.C.: Environmental Defense Publications.

Manly, W. L. (1894) 2004. *Death Valley in '49: Important Chapter of California Pioneer History.* San Jose, Calif.: Pacific Tree and Vine Co. Project Gutenberg Ebook, www.gutenberg.org/files/12236.

Mantua, N. J., S. R. Hare, Y. Zhang, J. M. Wallace, and R. C. Francis. 1997. A Pacific interdecadal climate oscillation with impacts on salmon production. *Bulletin of the American Meteorological Society* 78: 1069–79.

Masse, B. 1981. Prehistoric irrigation systems in the Salt River Valley, Arizona. *Science* 214 (23): 408–15.

Metropolitan Water District of Southern California. 1931. "Thirst," film produced by Gilliam and Reed, Hollywood.

Montgomery, D. R. 2003. *King of Fish: The Thousand-Year Run of Salmon.* Boulder, Colo.: Westview Press.

Moyle, P. B., M. P. Marchetti, J. Baldrige, and T. L. Taylor. 1998. Fish health and diversity: Justifying flows for a California stream. *Fisheries* (Bethesda) 23 (7): 6–15.

Moyle, P. B., and P. J. Randall. 1998. Evaluating the biotic integrity of watersheds in the Sierra Nevada, California. *Conservation Biology* 12: 1318–26.

Muir, John. (1901) 1991. *Our National Parks.* Reprint. San Francisco: Sierra Club Books.

Nichols, F. H., J. E. Cloern, S. N. Luoma, and D. H. Peterson. 1986. The modification of an estuary. *Science* 231: 567–73.

Nichols, F. H., J. K. Thompson, and L. E. Schemel. 1990. The remarkable invasion of San Francisco Bay (California, USA) by the Asian clam Potamocorbula amurensis, II. Displacement of a former community. *Marine Ecological Progress Series* 66: 95–101.

Plog, S. 1997. *Ancient peoples of the American southwest.* London: Thames and Hudson.

Powell, J. L. 2008. *Dead Pool: Lake Powell, Global Warming, and the Future of Water in the West.* Berkeley: University of California Press.

Powell, J. W. (1879) 1983. *Report on the Lands of the Arid Region of the United States, with a More Detailed Account of the Lands of Utah.* Facsimile of the 1879 edition. Boston, Mass.: Harvard Common Press.

Preston, W. 1981. *Vanishing Landscape: Land and Life in the Tulare Lake Basin.* Berkeley: University of California Press.

Reisner, M. 1993. *Cadillac Desert: The American West and Its Disappearing Water.* New York: Penguin Press.

San Francisco Estuary Project. 1992. *State of the Estuary: A Report on Conditions and Problems in the San Francisco Bay/Sacramento–San Joaquin Delta Estuary.* Oakland: Association of Bay Area Governments.

Swetnam, T. W., C. H. Baisan, A. C. Caprio, P. M. Brown, R. Touchan, R. S. Anderson, and D. J. Hallett. 2009. Multi-millennial fire history of the Giant Forest, Sequoia National Park, California. *Fire Ecology* 5 (3): 120–50.

Walker, T. 2009. Empty sound. *Sierra Magazine* (Dec.): 40–43.

Wilber, C. D. (1881) 2000. *The Great Valleys and Prairies of Nebraska and the Northwest.* Reprinted in *Wilber Republican,* Aug. 2.

Yoshiyama, R. M., E. R. Gerstung, F. W. Fisher, and P. B. Moyle. 2000. Chinook salmon in California's Central Valley: An assessment. *Fisheries* (Bethesda) 25 (2): 6–20.

CHAPTER 13. FUTURE CLIMATE CHANGE
AND THE AMERICAN WEST

Atwater, B. F. 1979. Ancient processes at the site of southern San Francisco Bay: Movement of the crust and changes in sea level. In *San Francisco Bay: The Urbanized Estuary,* edited by T. J. Conomos, 31–45. San Francisco: American Association for the Advancement of Science.

Barnett, T. P., and D. W. Pierce. 2008. When will Lake Mead go dry? *Water Resources Research* 44: W03201. doi:10.1029/2007WR006704.

Barnett, T., D. W. Pierce, H. G. Hidalgo, C. Bonfils, B. D. Santer, T. Das, G. Bala, A. W. Wood, T. Nozawa, A. A. Mirin, D. R. Cayan, and M. D. Dettinger. 2006. Human-induced changes in the hydrology of the western United States. *Science* 319: 1080–83.

Barnosky, A. 2010. *Heatstroke: Nature in an Age of Global Warming.* Washington, D.C.: Island Press.

Bates, B. C., Z. W. Kundzewicz, S. Wu, and J. P. Palutikof, eds. 2008. Climate change and water: Technical paper of the Intergovernmental Panel on Climate Change. Geneva: IPCC Secretariat.

Bolville, B. A., and P. R. Gent. 1998. The NCAR climate system model, version one. *Journal of Climate* 11: 1115–30.

Cayan, D. R., S. A. Kammerdiener, M. D. Dettinger, J. M. Caprio, and D. H. Peterson. 2001. Changes in the onset of spring in the western United States. *Bulletin of the American Meteorological Society* 82: 399–415.

Cayan, D., A. Luers, M. Hanemann, G. Franco, and B. Croes. 2006. *Climate Change Scenarios for California: An Overview.* Sacramento: California Climate Change Center.

Das, T., M. D. Dettinger, D. R. Cayan, and H. G. Hidalgo. 2011. Potential increase in floods in California's Sierra Nevada under future climate projections. *Climatic Change* 109: 71–94.

Dettinger, M. D. 2011. Climate change, atmospheric rivers, and floods in California: A multimodel analysis of storm frequency and magnitude changes. *Journal of the American Water Resources Association* 47 (3): 514–23.

Dettinger, M. D., D. R. Cayan, and K. T. Redmond. 2002. United States streamflow probabilities and uncertainties based on anticipated El Niño, water year 2003:

Experimental long-lead forecast bulletin. *Center for Land-Ocean-Atmosphere Studies* 11 (3): 46–52.

Gleick, P. H. 1987. Regional hydrologic consequences of increases in atmospheric carbon dioxide and other trace gases. *Climatic Change* 10 (2): 137–61.

———. 1988. The effects of future climatic changes on international water resources: The Colorado River, the United States, and Mexico. *Policy Science* 21: 23–39.

Gleick, P. H., and E. L. Chalecki. 1999. The impacts of climatic changes for water resources of the Colorado and Sacramento–San Joaquin river basins. *Journal of the American Water Resources Association* 35 (6): 1429–42.

Harte, J., A. Ostling, J. Green, and A. Kinzig. 2004. Climate change and extinction risk. *Nature* 427: 145–48.

Mote, P. W. 2003. Trends in snow water equivalent in the Pacific Northwest and their climatic causes. *Geophysical Research Letters* 30. doi10.1029/2003GL0172588.

Peterson, D. H., R. E. Smith, M. D. Dettinger, D. R. Cayan, and L. Riddle. 2000. An organized signal in snowmelt runoff in the western United States. *Journal of the American Water Resources Association* 36: 421–32.

Powell, J. L. 2008. *Dead Pool: Lake Powell, Global Warming, and the Future of Water in the West.* Berkeley: University of California Press.

Solomon, S., D. Qin, M. Manning, Z. Chen, M. Marquis, K. B. Averyt, M. Tignor, and H. I. Miller, eds. 2007. *Climate Change 2007: The Physical Science Basis. Contribution of Working Group I to the Fourth Assessment Report of the Intergovernmental Panel on Climate Change.* Cambridge, Engl.: Cambridge University Press.

Stewart, I., D. Cayan, and M. Dettinger. 2005. Changes towards earlier streamflow timing across western North America. *Journal of Climate* 18: 1136–55.

Swetnam, T. W. 1993. Fire history and climate change in Giant Sequoia groves. *Science* 262: 885.

Timmermann, A., J. Oberhuber, A. Bacher, M. Esch, M. Latif, and E. Roeckner. 1999. Increased El Niño frequency in a climate model forced by future greenhouse warming. *Nature* 398: 694–96.

U.S. Census Bureau. 2000. *U.S. Census, 2000.* www.census.gov.

Westerling, A. L., H. G. Hidalgo, D. R. Cayan, and T. W. Swetnam. 2006. Warming and earlier spring increase in western U.S. forest wildfire activity. *Science* 313: 940–43.

CHAPTER 14. WHAT THE PAST TELLS US
ABOUT TOMORROW

Barnett, T., D. W. Pierce, H. G. Hidalgo, C. Bonfils, B. D. Santer, T. Das, G. Bala, A. W. Wood, T. Nozawa, A. A. Mirin, D. R. Cayan, and M. D. Dettinger. 2006. Human-induced changes in the hydrology of the western United States. *Science* 319: 1080–83.

Das, T., M. D. Dettinger, D. R. Cayan, and H. G. Hidalgo. 2011. Potential increase in floods in California's Sierra Nevada under future climate projections. *Climatic Change* 109: 71–94.

Dettinger, M. D. 2011. Climate change, atmospheric rivers, and floods in California: A multimodel analysis of storm frequency and magnitude changes. *Journal of the American Water Resources Association* 47 (3): 514–23.

Dettinger, M. D., F. M. Ralph, T. Das, P. J. Neiman, and D. R. Cayan. 2011. Atmospheric rivers, floods, and the water resources of California. *Water* 3 (2): 445–78.

Gallaway, D. L., D. R. Jones, and S. E. Ingebritsen. 1999. Land subsidence in the United States. U.S. Geological Survey, Circular 1182.

Gleick, P. H. 2003. Global freshwater resources: Soft-path solutions for the 21st century. *Science* 302: 1524–28.

Gleick, P. H., J. Christian-Smith, and H. Cooley. 2011. Water-use efficiency and productivity: Rethinking the basin approach. *Water International* 36 (7): 784–89.

Gleick, P. H., and M. Heberger. 2012. The coming megadrought. *Scientific American* (Jan.): 8.

Glenn, E. P., K. W. Flessa, and M. J. Cohen. 2007. Just add water and the Colorado River still reaches the sea. *Environmental Management* 40: 1–6.

Hanak, E., J. Lund, A. Dinar, B. Gray, R. Howitt, J. Mount, P. Moyle, and B. Thompson. 2011. *Managing California's Water: From Conflict to Reconciliation.* San Francisco: Public Policy Institute of California.

Leopold, A., S. Flader, and J. B. Callicott. 1991. *The River of the Mother of God and Other Essays.* Madison: University of Wisconsin Press.

National Research Council. 2012. *Sustainable Water and Environmental Management in the California Bay-Delta.* Washington, D.C.: National Academies Press.

Overpeck, J., and B. Udall. 2010. Dry times ahead. *Science* 328: 1642–43.

Porter, K., L. Jones, D. Cox, et al. 2011. Overview of the ARkStorm scenario. U.S. Geological Survey, Open-File Report 2010–1312.

Postel, S. 1992. *Last Oasis: Facing Water Scarcity.* New York: Norton Press.

Reisner, M. 1993. *Cadillac Desert: The American West and Its Disappearing Water.* New York: Penguin Press.

Westerling, A. L., H. G. Hidalgo, D. R. Cayan, and T. W. Swetnam. 2006. Warming and earlier spring increase in western U.S. forest wildfire activity. *Science* 313: 940–43.

INDEX

Swetnam, Thomas, 133–34, 195
Syringa vulgaris, 190–91

Tahoe, Lake: drought of 1976–77 level, 21; Dust Bowl level, 44; Native Americans living around, 100; Neoglacial level, 113; Pyramid Lake linked to, 100; submerged trees, 97–100, **98**
Tambora, Mt., 163
Taylor Grazing Act, 198
Tenaya Lake, 130
Thalassionema nitzschioides, 106
"Thirst" (film), 178–79
tree-ring studies: age resolution and, 79; climate history clues from, 64–67, 68–69, 77; evidence of Colorado River flow fluctuations from, 67, 150–51; evidence of Medieval drought climate extremes from, 131–33, 135; evidence of Medieval drought wildfires from, 133–34, **133**, **134**; solar cycles and, 162
trees: dying during drought of 1976–77, 22–23; killed by bark-beetle disease, 22, 46, 194; submerged in lakes/rivers, 74, 97–100, **98**, 130–31, **131**. *See also specific trees*
Truckee River, 21, 44, 100, 113
Tulare Lake: as candidate for restoration, 217; early Holocene level, 88; history of, 119, 186–87; Little Ice Age level, 150; mid-Holocene level, 103–4; Neoglacial level, 113; Younger Dryas level, 91
Tule River, 113, 187
Tuolumne River, 179, 193–94
Twain, Mark, 2

Udall, Bradley, 212
upwelling: defined, 52; during La Niña, 53; during Medieval drought, 137; during mid-Holocene, 105–7; salmon populations and, 56–57, 182; solar cycles and, 161
Urey, Harold, 167
U.S. Geological Survey, Atmospheric River 1000 Storm scenario of, 210–11
U.S. Southwest. *See* Southwest

Utah, floods of 1861–62 in, 37
Uto-Aztecan people, 108, 109–10

varves, defined, 71
vegetation: altered by mid-Holocene drought, 104; at beginning of Holocene, 92–94; flowering earlier, 190–91, 197; salt-tolerant, 104, 125–27, **128**. *See also* plant communities; trees
Verde River, 25, 38, 123
volcanic eruptions: climate change and, 162–63; determining age of sediments containing ash from, 78–79, 93; during Medieval period, 132; megafloods and, 149

Walker, Gilbert, 51
Walker, Phillip, 137
Walker Circulation, 51
Walker Lake, 85
Walker River, 130, **131**
warmer temperatures, 190–203; causes of, 202; coastal impacts of, 200; Colorado River basin and, 195–96; ecosystem impacts of, 196–98; increased flood risk, 199–200; increased risk of dam failures, 200–202; indicators of, 190–91, 193–95, **195**; precipitation patterns altered by, 191–92; records of, 191, **192**; risk of Southwest dust bowl, 198. *See also* climate change
water conservation: difficulty of getting people to adopt, 212; drought of 1976–77 as inspiring interest in, 24; importance of, in future, 189, 203; lessons from Australian drought on, 213; standards and pricing encouraging, 220
water development: bringing water to Southern California, 44–46, 103, 177–80, **177**, **180**; dam removal to reverse ecosystem damage from, 217–18; by early Native populations, 123–24, 136; population growth made possible by, 8; potential for drought obscured by, 7; reversing environmental damage from, 215–17, **216**; transformation of Central Valley by, 180, 185–87, 208
water footprint, 213–14

TEXT
11/14 Adobe Garamond Premier Pro
DISPLAY
Adobe Garamond Premier Pro
COMPOSITOR
IDS
PRINTER AND BINDER
Maple Press